Medical Intelligence Unit

# The Etiopathogenesis of Coronary Heart Disease: A Heretical Theory Based on Morphology

## Second Edition

Giorgio Baroldi, M.D., Ph.D., FACC, FESC
Department of Pathological Anatomy
University of Milan
Institute of Clinical Physiology
National Research Council
Pisa and Milan, Italy

Malcolm D. Silver, M.D. Ph.D., FRCPA, FRCPC
Department of Laboratory Medicine and Pathobiology
University of Toronto
Toronto, Ontario, Canada

CRC Press
Taylor & Francis Group
Boca Raton London New York

CRC Press is an imprint of the
Taylor & Francis Group, an **informa** business

# The Etiopathogenesis of Coronary Heart Disease: A Heretical Theory Based on Morphology

## Second Edition

### Medical Intelligence Unit

CRC Press
Taylor & Francis Group
6000 Broken Sound Parkway NW, Suite 300
Boca Raton, FL 33487-2742

First issued in hardback 2019



ISBN 13: 978-1-58706-076-2 (hbk)



**Visit the Taylor & Francis Web site at**
**http://www.taylorandfrancis.com**

**and the CRC Press Web site at**
**http://www.crcpress.com**



### Library of Congress Cataloging-in-Publication Data

Baroldi, Giorgio.
The etiopathogenesis of coronary heart disease : a heretical theory based on morphology / Giorgio Baroldi, Malcolm D. Silver.-- 2nd ed.
p. ; cm. -- (Medical intelligence unit)
Rev. ed. of: Sudden death in ischemic heart disease / Giorgio Baroldi, Malcolm D. Silver. c1995.
Includes bibliographical references and index.
ISBN 1-58706-076-0
1. Coronary heart disease--Etiology. I. Silver, Malcolm D. II. Baroldi, Giorgio. Sudden death in ischemic heart disease. III. Title. IV. Series: Medical intelligence unit (Unnumbered : 2003)
[DNLM: 1. Coronary Disease--etiology. 2. Coronary Disease--physiopathology. WG 300 B264e 2003]
RC685.C6B285 2003
616.1'23071--dc22

2003027906

# Dedication

To our critics and those we have criticized

# CONTENTS

# PREFACE

Heresy from the Greek (αιρεσισ) means "choice" and semantically indicates an opinion or doctrine at variance with what is currently orthodox. A heresy is mainly considered an error by the orthodox and one that must be eradicated to preserve current thought and their dominion. Nevertheless, human knowledge, including science, progresses through paradoxes or heresies. It takes time for the latter to be proven wrong or be accepted. For example, Aristarchus of Samos, about 270 B.C., anticipated the heliocentric theory which Copernicus presented in his 1543 treatise "De revolutionibus orbium coelestium" and which was demonstrated by Galileo in his "Dialogo sopra i due massimi sistemi del mondo" of 1632. In the interim, between Copernicus and Galileo and even afterwords any heretic fostering these concepts faced the power of the Inquisition, then defender of academic dogma. Once a heretic was burned, today silence replaces the pyre presenting an intellectual barrier to discussion and curtailing scientific criticism of methods and findings. But, if a heresy is based on solid facts sooner or later that barrier will fall. A heretic has the moral duty to advertise the new concepts adopting any means available to capture the attention of those not yet incarcerated by dogma; mainly young scientists or practictioners in training and some older individuals, who maintain an open, critical mind seeking only the truth, and being aware that dogma prevents cultural evolution.

In proposing a heretical scientific opinion one must define the natural history, step-by-step and fact by fact, rationalizing the construction of the opinion and its right to exist. In doing so, the need is to validate any methodologic approach, which, in our case, is morphology. Before the exponential growth of clinical technology in the few past decades, pathology was the gold standard for any diagnostic and pathophysiologic interpretation; a standard established in the late XVI century when Morgagni understood the need to see within a body what caused illness and death. At present, one sees a reduction in autopsy requests by clinicians due to their presumption that clinical imaging can substitute for morphological findings. However, demonstrating dysfunction or altered morphology by clinical techniques cannot necessarily establish its cause and pathogenic mechanism. No matter how important it is to show a dysfunction, the need is to see what the cellular alterations are; having in mind that morphologic techniques have also improved dramatically. The point is to perform an autopsy immediately after death, and to have—as for organ donations—fresh tissue for electron microscopy, immunohistochemistry, tissue biochemistry and molecular biology. Then, a clinical-pathological collaboration, based on morpho-functional control of any imaging obtained in vivo, becomes compulsory. In this book, as far as cardiology is concerned, the reader will find many examples indicating how dangerous is the present concept that only clinical imaging can ratify hypotheses and that morphologic support, is unnecessary.

Another point to be stressed is that both clinicians and pathologists study patients in whom a disease may have started long before, and thus have little possibility of documenting how it started and progressed. Furthermore, many clinical techniques are invasive, so cannot be employed for a more precise and correct study of the general population; such monitoring being unrealistic at present. In contrast, a pathologist may compare any variable in the general population, either in patients with different diseases or in normal subjects dying by accident. As yet, experimental models cannot fully reproduce the natural history of most degenerative diseases.

A final criticism concerns the current bad habit of authors, who only review the literature of the past few years. Such writings reflect dogma a la mode, and omit essential earlier contributions, particularly any that are at variance with current beliefs.

With these thoughts in mind, we write the second edition of this book according to the historical sequence of our studies by research protocols, each one being the logical continuation of previous findings. This is how science progresses. We begin with the study of coronary collaterals and end with that of hearts from patients with congestive heart failure. In contrast to the current belief our data show that:

1. In human hearts coronary collaterals exist and compensate severe atherosclerotic stenoses. At the initial presentation of coronary heart disease in apparenlty normal subjects, severe single or multiple coronary stenoses preexisted in absence of symptoms and signs.
2. Atherosclerosis is due to increased hemodynamic stress on the vessel wall secondary to recurrent or stable regional myocardial dysfunction and increased peripheral resistance. The latter is due to extravascular compression of the intramural vessels within an asynergic zone of myocardium. All changes seen at the plaque level (hemorrhage, rupture, thrombosis) are secondary phenomena to obstruction of flow.
3. Coronary atherosclerosis in man has a different history and structure from atherosclerosis following hypercholesterol diet or familial hypercholesterolemia.
4. Myocardial necrosis in coronary heart disease is not a pool of different myocardial changes (coagulation necrosis, contraction band necrosis or apoptosis) due to ischemia.[1] It is a collection of distinct forms of myocardial injury each with its own etiopathogenesis: blood flow reduction for *infarct necrosis,* catecholamine myonecrosis for *contraction band necrosis* linked to malignant arrhythmia/ventricular fibrillation and colliquative myocytolysis due to a non ischemic metabolic disorder of myocardial cells ending in *congestive heart failure.*
5. Morphologic data support the hypothesis that coronary heart disease is more an adrenergic stress-dependent disease than a hydraulic problem.

*Giorgio Baroldi*
*Malcolm D. Silver*

**Reference**

1. Thygesen K, Alpert JS. Myocardial infarction redefined. A consensus document of the joint European Society of Cardiology/American College of Cardiology Committee for the redefinition of myocardial infarction. J Am Coll Cardiol 2000; 36:659.

# Acknowledgements

We thank all our coworkers in the different research protocols and in particular, statisticians Fabio Mariani, Gabriella Giuliani and Marina Parolini and our secretaries Elisabetta Spagnolo and Diana Houghton. Grants were received from the Ontario Heart and Stroke Foundation, Targeted project FATMA, National Research Council, Rome. We are grateful to Fiorella Baroldi and Meredith Silver for continuing understanding and support.

# Chapter 1

# Plastic Cast Study of Coronary Vessels

Coronary heart disease (CHD) becomes epidemic in any society that transforms itself from an agricultural one to an industrial-technological system with an associated behavioural alteration in the populace's diet, stress and consumerism. Despite risk factors prevention and many new therapeutic approaches, this disease remains the first cause of morbidity and death in such societies, a primacy that carries enormous implications particularly as CHD often affects people at the peak of their experience and productivity. Furthermore, cardiovascular diseases associated with atherosclerosis will be one of the main health care problems of this millenium in which an increasing mean age plus a reduction of mortality due to treatment of acute vascular events will result in an excessive number of individuals with chronic cardiovascular conditions who need continuous assistance and rehabilitation. This is a human and social cost that will be hard to sustain.

## Definition of Coronary Heart Disease

The present definition is an up-to-date version of persisting concepts from the earliest reports of the first few cases of acute myocardial infarction (Hammer, 1878; Herrick, 1912, 1919), the cause of which was considered to be an occlusive thrombus and "explosion" or rupture of an atherosclerotic plaque with the latter, proposed to cause the occlusive thrombus (Chapman, 1974). More recently we learned that acute coronary syndromes should be interpreted as follows: "Evidence from serial coronary arteriography and that obtained after reperfusion by thrombolysis, at operation during acute coronary syndromes and from postmortem arteriography have also confirmed the importance of plaque disruption and thrombosis. Indeed, these acute or subacute changes in coronary arterial anatomy appear to be the most frequent cause of all the acute coronary syndromes including unstable angina, myocardial infarction and ischemic sudden death. If we accept the premise that all three acute coronary syndromes may evolve from acute plaque disruption followed by thrombosis or spasm or both, we can construct an unifying theory" (Gorlin et al, 1986); and "recent advances in cardiovascular molecular biology, coronary diagnostic technique and cardiac therapeutics have opened windows of opportunity to study and modify the factors leading to plaque rupture. The local modification of gene expression to alter plaque composition and to elucidate and subsequently inhibit the prothrombotic and fibrinolytic defects that promote coronary thrombosis may in future prevent plaque rupture and its consequence" (Mac Isaac et al, 1993). A prevention optimistically foreseen for the first years of the new millenium (Braunwald, 1997).

This "hydraulic" theory, blessed by molecular biology findings in the plaque (Libby, 1995), is mainly based on our inability to image coronary collaterals angiographically (Helfant et al, 1970). In fact, if functioning collaterals exist the whole hydraulic theory needs reevaluation.

## Origin of Heresy

The history of our study (and heresy) began in 1952 when in the old basement of the Institute of Pathological Anatomy of the University of Milan one of us (GB) began injecting the arteries of the heart to investigate whether coronary collaterals—or anastomoses—exist. At that time, their existence was questioned and their significance in humans practically unknown, a crucial point in understanding the etiopathogenesis of CHD. Apart from some experimental data in dogs (Gregg, 1950), human postmortem study were mainly negative (Blumgart et al, 1940) or inconsistent in demonstrating presence and function of collaterals in normal and pathological hearts. The prevailing idea was (and remains) that they are absent or unable to function.

A complete review of the pertinent literature was reported in a preliminary paper (Baroldi et al, 1956) which was followed by a monograph (Baroldi et al, 1967) subsequently cited as a key reference for coronary arteriography (Higgins, 1980). Afterwards, as far as we know, apart from one significant further pathological contribution (Zamir et al, 1985) all data on human collaterals have been based on observations derived from coronary cineangiography.

## Plastic Cast Study of the Coronary Vessels

The anatomy of the coronary vascular system was investigated with particular reference to collaterals or anastomoses defined as channels joining branches of different arteries. This study included: 50 normal adult individuals who died by accident; 19 normal hearts from noncardiovascular disease patients; 10 children from newborn to 10 years of age all with normal hearts; 25 atrophic adult hearts and 48 with left or right or total ventricular hypertrophy of the hearts, all having normal coronary arteries; 18 patients with chronic anemia and/or hypoxic diseases but a normal heart; 217 patients with obstructive lesions of any degree of the coronary arteries: 70 cases of out-of-hospital sudden/unexpected coronary death and 147 hospitalized patients; of the latter, 47 died of acute myocardial infarction, 21 associated with extensive fibrosis and 100 had a normal myocardium or minimal myocardial fibrosis and died of noncardiac causes. All cases were selected according to the same criteria mentioned below.

Plastic material (Geon latex 756 or Neoprene 842 A) was injected into coronary arteries under 130-200 mmHg pressure through the aorta after hermetically closing the aortic valve. Penetration of the plastic material was enhanced by rhythmic compression of the aortic bulb and a light massage of the suspended heart. The injected material was solidified by placing the hearts in 10% formalin at 40-50° C for 48-72 hours. Before corroding the organ in concentrated hydrochloric acid solution, to allow formation of accurate tridimensional casts of the coronary tree (Fig. 1), several myocardial samples from different cardiac areas were secured for histology. The lack of shrinkage during solidification permitted a correct evaluation of luminal diameter, its reduction in case of stenosis and gave a tridimensional view of all arterial vessels, to a diameter of about 20 μm. The diameter of vessels measured in these casts was considered the diameter in maximal dilatation since they were injected and fixed under pressure.

The arterial collateral circulation was estimated by an anastomotic index (AI) formulated as follows: AI = Max Ø +(AV Ø x Frequency)/100 in which Max Ø was the diameter in microns of the largest anastomotic vessel found; AV Ø was the average of the diameters of the larger anastomoses greater than 100 μm; Frequency was the number of anastomoses greater than 100 μm found in any heart. In this respect an average of 40 such anastomoses were usual in a normal heart, providing an arbitrary index of 1. Collaterals were distinguished as homocoronary, when they connected branches of the same coronary artery or intercoronary, when branches joined different coronary arteries. Extracardiac collaterals ran between coronary branches and other adjacent arterial systems e.g., bronchial arteries (Moberg, 1968).

Finally, by injection of the coronary sinus or heart cavities, the coronary venous system (coronary sinus system and anterior cardiac venous system) was studied in 74 cases. Also, the arterioluminal and venoluminal vessels i.e., connections between coronary arteries or veins and cardiac chambers were investigated in 48 cases and extracardiac arterial connections in 13 cases (Baroldi et al, 1967).

Reviewing the casts, the following patterns of coronary artery distribution were defined: In type I (77%) the right coronary artery (RCA) gave rise to the posterior descending branch. According to the length of the RCA three subtypes were recognized. In type I a (5%) the RCA ended as soon as it became the posterior descending branch without significant ramifications on the posterior left ventricle; in type I b (55%) the RCA vascularized half of the posterior left ventricle and in type I c (17%) all of the posterior left ventricle was nourished by the RCA which ended at the left cardiac margin. In type II (8%) distribution the posterior descending branch originated from the left circumflex artery: and in type III (15%) two posterior descending branches existed, one each arising from the left circumflex and right coronary artery respectively. (Fig. 1). A third coronary artery (conus artery or arteria coronaria accessoria dextra) was observed in 46% of cases; occasionally, it was double (8% of cases) or triple (1%).

## *Collaterals*

Both homo- and intercoronary collaterals, form an extensive network in all regions of the normal heart (Fig. 2). The collaterals join adjacent arterial vessels at different levels along their course and within the whole thickness of the cardiac wall. They have a characteristic finely coiled appearance which seems related to their course, parallel to the line of cardiac muscle contraction.

Human collateral vessels are intramural in location and have a capillary-like wall. However, two hearts, one normal and one with severe coronary artery stenoses showed respectively one capillary-like superficial collateral of 100 μm and, large anatomoses (500-1000 μm or more in diameter) in the interstitial tissue between visceral pericardium and myocardium. The latter vessels had an abortive thin media with a fragmented, ill-defined elastic lamina. The diameter of collaterals in normal hearts ranged from less than 20 to 350 μm with an anastomotic index (see above) from 3.4 to 6.2 (mean 4.7). In atrophic hearts with normal coronary arteries the index was 2.5-6.4 mean 3.7; amongst hypertrophied hearts with normal coronary arteries the figures were 4.5-14.0, mean 7.4. Normal hearts from subjects with hypoxic conditions (e.g. chronic anemia) had an index of 9-19 (mean 12) (Table 1; Fig. 3).

In the presence of atherosclerotic stenosis with a lumen/diameter reduction greater than 70% there was a dramatic increase in the diameter and length of collaterals which might exceed 1000 μm in diameter and be several centimeters long. Any severely obstructive coronary artery lesion, even multiple ones, was always found associated with enlarged collaterals. These could be intercoronary or homocoronary and joined the vessels bypassing the obstruction(s) (satellite anastomoses) (Fig. 3-4). The anastomotic index in these instances ranged from 5 to 33 with a mean value of 16 associated with a single stenosis and 22 in multiple severe stenoses. The presence or absence of a myocardial infarct was independent of the number of enlarged collaterals. In other words, hearts with the same degree of obstructive coronary damage could show a similar pattern of collateral enlargement unrelated to their myocardial status, whether normal or affected by an acute infarction or showing fibrosis of any extent. Similarly, sudden and unexpected death cases did not show any divergency with respect to the number and size of collaterals. According to the anastomotic index, collateral enlargement increased proportionally with an increasing degree of stenosis and the number of severely obstructed coronary vessels in each heart. The peculiar satellite anastomotic network seen at the site of advanced plaques will be described later.

## *Collateral Function According Our Data*

In contrast to results obtained by others (Blumgart et al, 1940)), the tridimensional coronary casts showed that homo-and intercoronary collaterals normally exist in the human heart. They are demonstrable at birth and may work in infancy. For example, in cases with an anomalous origin of a coronary artery from the pulmonary artery, left-to-right shunts caused by a pressure gradient between aorta and pulmonary arteries occur via collaterals. The casts also indicate that in three circumstances homo- and intercoronary collateral vessels increase in size, namely in cardiac hypertrophy, in chronic hypoxic diseases with normal coronary arteries and in the presence of critical coronary stenoses. Only in atrophic hearts with normal coronary arteries are the diameters of collaterals less than normal (Table 1). Therefore, anastomotic channels may enlarge by dilatation and/or hyperplasia of cellular-tissue components of the vessel wall in different conditions, for different causes and with different functional meanings. We have already described the capillary-like composition of the wall of intramural collateral vessels (Fig. 4), with minor and insignificant structural changes occurring even at maximal enlargement, when anastomoses are extramural. We wonder if the term "angiogenesis" used to describe this condition is correct since, in the normal myocardium, there is no proof of new vessel formation as occurs, for instance, in granulation tissue. In our opinion the latter rarely forms in the myocardium. We believe that the repair of an infarct or other types of myocardial necroses is accomplished by collagenization of sarcolemmal tubes without new vessel formation (Baroldi et al, 1975). Granulation tissue forms and angiogenesis occurs along the course of a traumatic wound of the myocardium or following implantation of the internal mammary artery or Vineberg's operation, a now abandoned surgical attempt to revascularize the myocardium. Following the latter procedure newly formed anastomotic channels may participate in flow redistribution. However, when implantation is performed in a collateral dependent zone with severe stenosis or occlusion of the related artery, the amount and direction of flow in new anastomoses becomes questionable because of competing flow from preexisting collaterals. Lysamine dye injected into the implanted artery remains limited to the immediate area of the implant, unless the main coronary branch is cross-clamped distal to the previous occlusion. In that case dye distribution approximates the myocardial zone dependent upon the main occluded branch (Mantini et al, 1968). Angiogenesis at the site of an implanted internal mammary artery which provides modest nutritive blood-flow to a collateral-dependent region has been considered a model to promote neovascularization suggesting the possible role of a local angiogenic factor that enhances new collateral formation (Unger et al, 1990). We believe that is a very unlikely phenomenon. In general, rather than "angiogenesis" (new vessel formation) we should speak of angiohyperplasia producing tridimensional enlargement of preexisting normal vessels by hyperplasia of their wall components as was observed in both arterial and venous extra- and intramural vessels in hypertrophy of the human heart (Baroldi et al, 1967) and in experimental right ventricular hypertrophy (Farb et al, 1993). A similar angiohyperplasia of intramural arterial vessels is seen in chronic hypoxia (see discussion below on myocardial infarct necrosis).

Different mechanisms that trigger collateral angiohyperplasia probably exist. In myocardial hypertrophy/hypoxia an increased oxygen demand (Scheel et al, 1985, 1990) can induce a generalized increase in vessel size by stimulating endothelial growth factors which exist in myocytes (Speir et al, 1988; Weiner et al, 1989; Sasaki et al, 1989; Sasayama et al, 1992). Apparently, collateral growth is also enhanced by heparin (Fujita et al, 1988; Carroll et al, 1993; Quyyumi et al, 1993). The selective increase of satellite collaterals in coronary heart disease suggests that a pressure gradient, or increased collateral flow velocity (Flynn et al, 1993) rather than ischemia, are the main stimuli, through endothelial cell growth factor (D'Amore et al, 1987, Rajanayagam et al, 2000); thus explaining the relationship between collateral flow and intracoronary growth factor concentration in patients with stenosis in different vessels shown at angioplasty (Fleish et al, 1999). In fact, in the same ischemic zone, a single enlarged collateral may coexist with normal ones; in other words not all collaterals enlarge in the whole ischemic area as they should if ischemia were the main stimulus for their development (Fig. 4). A fact which contradicts the concept that ischemia, per se, is a collateral-genic factor (Chilian et al, 1990).

Two types of collateral development can be distinguished. One is diffuse and affects the whole collateral system associated with an increase in diameter and length of the total intramural system. This is seen in myocardial hypertrophy or hypoxic states. The other is a regional development being located in relation to a specific lumen reduction of a main arterial vessel (satellite anastomoses). Accordingly, two types of function for collaterals can be considered:

1. In diffuse enlargement, in the presence of normal coronary arteries, the function is mainly nutritional the collaterals forming part of the terminal vascular bed. Their capillary-like structure, their spatial disposition parallel to the plane of contraction of cardiac muscles with corkscrew adaptation (Fig. 2) and with a systolic flow as demonstrated in cardiac capillaries (Tillmans et al, 1974), indicate a capillary-like function, i.e., the delivery of nutrient substances to the myocardium. Their enlargement parallel with all intramural vascular structures balances an increased demand of the hypertrophied or hypoxic myocardium.
   The inclusion of the collateral system in the terminal vascular bed suggests some considerations in relation to the delivery of oxygen and other substances from capillary to myocell. The geometrical model of one capillary to one cell (Wearn et al, 1928) or four equally spaced capillaries per myocell (Ludwig, 1971) should be reconsidered. In reality we deal with a more extensive surface and ubiquitous disposition of the terminal network, collateral vessels surrounding each myocell in all directions. This complex terminal system seems more appropriate in face of a wide range of metabolic demands, even in markedly hypertrophied hearts. The lack of relation between infarct size and heart weight speaks against the supposed relative ischemia of hypertrophied myocardium, particularly in cor pulmonale. It does work for many years (Baroldi, 1971).
2. Redistribution of nutritional flow is the second function of collaterals in the presence of critical coronary obstruction(s) (satellite anastomoses). The enlargement and extent of satellite pattern are proportional to the number of severe obstructions and appear to supply nutritional blood to the dependent "ischemic" territory.

A fact to stress is the high variability of the anastomotic index in the different groups shown by the coronary plastic cast study (Table 1). Another is the different pattern of collateralization found in the presence of coronary obstruction of comparable location and degree. Occlusion at the same site of the same main coronary arterial vessel may be "compensated" by relatively few, very enlarged anastomoses, easily seen by cineangiography, or by numerous, relatively small (100-300 μm), anastomoses that are not, or are poorly demonstrated by cineangiography (Fig. 4). Despite assertions that angiographic techniques permit one to see intramural vessels up to 100 μm in diameter (Gensini et al, 1969), a comparison between plastic casts of coronary arteries and their cineangiographic images clearly shows the inadequacy of the latter in visualizing intramural vessels.

Two factors may be responsible for the high variability of collateral patterns found in coronary heart disease seen either at postmortem or in vivo. The first is a progression of coronary atherosclerosis in the whole extramural coronary arterial system. Flow redistribution and related anatomical changes would depend on the chronological development and location of plaques in different coronary arteries or their branches. A second factor is infarct necrosis. In the latter condition, proportional to its size, postmortem coronary injection of radiopaque or plastic material fails to fill intramural vessels in an infarcted zone. Stretching and compression of infarcted myocardium by intraventricular pressure and subsequent thrombosis of intramural vessels within necrotic tissue lead to an intramural avascular area (see infarct necrosis) where only extramural branches can be seen. In a healed infarct the histologic picture is of hyaline, dense and avascular connective tissue, sometimes crossed and often surrounded by giant capillaries (angiomatous plexus). The latter are, in fact, already visible during repair. The disappearance of all intramural arterial vessels, including collaterals, suggests that surviving anastomoses may further enlarge since the pressure gradient between the stenotic infarct-related artery and adjacent vascular territories persists. These highly enlarged collaterals are those seen by angiography. The plaque satellite anastomotic network in advanced plaque will be discussed in the Chapter 5.

## *Angiographic Imaging of Collaterals in Vivo*

In many angiographic studies of patients undergoing aortocoronary bypass grafting, angioplasty or intracoronary thrombolysis following an acute myocardial infarction, the presence of collaterals has been investigated and correlated with other functional variables. An increased frequency of collaterals was seen in relation to the number of main vessels with severe stenosis (> 50%); however, both the number of stenoses and the presence or absence of collaterals did not correlate with hemodynamic abnormalities, e.g., left ventricular end-diastolic pressure and cardiac index. Ventriculographic alterations were associated with an increasing number of diseased vessels but not with the presence of collaterals, suggesting the lack of their protection. (Helfant et al, 1970; Bodenheimer et al, 1977). A flow restriction was observed in collateral dependent myocardium in patients with complete occlusion of the left anterior descending branch (Arani et al, 1984). Other patients undergoing aorto-coronary bypass grafting with total coronary artery occlusion associated with collaterals, showed hemodynamic changes, i.e., post-stenotic coronary pressure and graft flow hyperemia, that simulated those of a 90% coronary stenosis without any collaterals. On the other hand, left ventricular asynergy increased with the severity of coronary obstruction without relation to angiographically significant collaterals: 63% of hypo- or akinetic wall segments in chronic coronary occlusion with collaterals versus 45% with < 80% stenosis, 52% with 80-90% stenosis and 57% with 91-99% stenosis without collaterals; (Flameng et al, 1978).

Patients with angina pectoris and persistent occlusion of a major coronary artery but without previous myocardial infarction have, by positron emission tomography, a similar regional myocardial blood flow and both oxidative metabolism and glucose uptake in collateral-dependent and remote myocardial segments with normal wall motion. In contrast, a lessened myocardial blood flow, reduced oxidative metabolism and higher glucose uptake were observed in dysfunctioning collateral-dependent myocardial segments versus normal, remote, ones. However, when collateral-dependent segments with and without abnormal wall motion were compared, no differences in blood flow were found. After intravenous dipyridamole, collateral-dependent myocardial blood flow greatly increased in segments with normal wall motion, while the increase was minimal if associated with asynergy. A functional follow-up in 12 patients undergoing percutaneous transluminal coronary angioplasty (8 cases) or coronary artery bypass surgery (4 cases) revealed adequate revascularization in 11 (reocclusion in one after successful angioplasty). In all, including the reoccluded subject, regional wall motion improved. (Vanoverschelde et al, 1993). In another positron emission tomography study, patients with stable exertional angina, normal ventricular function and chronic occlusion of a major subepicardial artery, opacified via intramyocardial collateral flow, myocardial blood flow at rest equalled that in normal volunteers in areas of normal myocardium and in myocardial collateral-perfused areas of the patients. However, following dipyridamole the increase in flow in collateral-perfused areas was only one half that in normal areas (Mc Falls et al, 1993).

The many clinical studies cited and numerous editorials (Cohen, 1978; Gregg et al, 1980; Topol, 1991; Sasayama et al, 1992) emphasize uncertainties about the functional role of coronary collaterals in coronary heart disease. In fact, other papers document the role of preexisting collaterals in preventing postinfarct left ventricular aneurysm formation (Forman et al, 1986; Habib et al, 1991) despite their inability to improve ventricular function (Hirai et al, 1989). Amongst patients who experience unsuccessful thrombolysis, in the presence of collateral vessels at the onset of a myocardial infarct, other authors found a limitation of infarct size as assessed enzymatically (Habib et al, 1991) and improved ventricular function determined by left ventricular ejection fraction (Williams et al, 1976; Nohara et al, 1983; Rogers et al, 1984; Schwartz et al, 1985; Saito et al, 1985; Habib et al, 1991). Such improved ventricular function was also observed after late thrombolytic therapy (within 12 hours) but not with nitroglycerin intracoronary infusion (Rentrop et al, 1989). The successful recanalization by intracoronary thrombolysis in acute infarct patients without demonstrable collaterals did not improve left ventricular function (Rogers et al, 1984; Saito et al, 1985).

After a long period of "aggressive cardiology" in this "catheter" era, it is time to review the formulated hypotheses on the etiopathogenesis of ischemic heart disease. Three considerations are pertinent when the morpho-functional significance of structural parameters are reconsidered. First, as already stated, clinicians and pathologists examine only what has been selected by already advanced disease. Second, there is difficulty distinguishing primary from secondary events in the course of a disease. Third, to clinically monitor ischemic heart disease phenomena before they happen, when they begin, during their course to the end and to examine histologically the heart has been reported only once in the literature (Baroldi et al, 1990); nor do we have an experimental model of this disease. Each of these points is important when one considers cause-effect relationships in respect of a still incomplete natural history.

In our opinion, the phenomena associated with coronary heart disease are not only a hemodynamic problem but many other factors are involved. The aim is to review the relationship of the variables recognized in its history and to discriminate fact from fiction. We will proceed step by step, with the same sequence we did in our studies, having in mind Wilson's observation (1952): If one doubts the necessity of controls, reflect on the statement: it has been conclusively demonstrated by hundreds of experiments that the beating of tom-toms will restore the sun after an eclipse".

## Conclusions from Coronary Plastic Casts Study

In coronary cineangiography, the diagnostic likelihood of myocardial ischemia is based on the percentage of lumen reduction found in the coronary system. Consequently, all effort is oriented to perfecting the evaluation of stenoses and relieving them. Allowing for technical pitfalls and real difficulties in obtaining correct measurements, the effect on coronary flow of a lumen reduction depends on many variables. It is directly proportional to the fourth power of the radius of the lumen and is inversely proportional to fluid viscosity and the length of the tube. In particular, the pressure difference across a constricted segment is determined by pressure on the upstream side, the resistance to flow through the constricted segment and the peripheral resistance of the arteries and vascular bed distal to the constriction (Gregg, 1950). In artificial systems a maximal increase in peripheral resistance may greatly reduce flow, even in the presence of a normal lumen. In animals, prediction of flow reduction in relation to the degree of stenosis becomes more difficult because of the response of the peripheral vascular bed. In the coronary bed peripheral resistance is relatively high and generally sizeable reductions in lumen are needed before inflow diminishes (Gregg, 1950). Flow reduction, therefore, is difficult to evaluate particularly if we include such other variables as vessel tone, collateral flow and myocardial contractility. At present, the prevailing concept is that coronary arteries are, physiologically speaking, end arteries and even if demonstrated anatomically, collaterals cannot protect the myocardium from an acute ischemic event. In this viewpoint, nutrient flow to the distal myocardium totally depends upon the amount of blood that passes a stenosed lumen of an extramural coronary artery. The rationale is that even a pin-point coronary artery residual lumen (90% and 99% lumen reduction of a vessel with a diameter of 4 mm corresponds to a residual lumen of 400 μm or 40 μm respectively) allows a satisfactory flow that can be put in jeopardy only by (a) vessel occlusion, leading to myocardial infarction, or (b) increased metabolic demand, causing angina pectoris. Overall, this proposal proved difficult to accept when we saw the first tridimensional view of highly enlarged collateral vessels, satellite to a stenosis. The total caliber of these collaterals greatly exceeded that of the residual stenosed lumen. We realized that an atherosclerotic stenosing plaque is not a static element but is the site of integrated dynamic and biomolecular events that must be considered. Collaterals are part of the normal vascular system of the human heart and are too numerous to be neglected. They are present at birth, within the whole thickness of the cardiac wall and in any cardiac region. Furthermore, extracardiac collaterals connect the coronary with adjacent arterial systems. It must be stressed that in man collaterals connecting subepicardial coronary arterial branches are uncommon; rather, the inter-homocoronary collateral system is mainly intramural.

The basic question is whether blood flow redistribution through collaterals may, or may not, prevent ischemia in the vascular territory of a stenosed or occluded coronary artery. From postmortem tridimensional casts one can say that an increase in collateral size appears to be a sign of increased function. A successful compensatory mechanism can be deduced by the following observations in our cast study:

1. Old total coronary occlusions were found in "healthy" subjects who died from accident but had no ischemic heart disease clinically, nor any significant myocardial fibrosis.
2. Patients who died from noncardiac diseases, had no cardiac disorder and/or extensive myocardial fibrosis despite having occlusions or multiple severe coronary artery stenoses. Of 217 consecutive coronary atherosclerotic patients, 46 had mild and 171 severe lumen reduction of at least one main coronary artery. An infarct was documented in 5 (10%) and 68 (40%) of these cases respectively (Baroldi et al, 1967). In particular, a high frequency of old severe stenosis (≥90% lumen diameter) was observed without histologic evidence of a myocardial infarction or extensive myocardial fibrosis.

Additional evidence comes from cases of aortitis with bilateral coronary ostial occlusion without ischemic heart disease and from the observation that new total coronary occlusions are demonstrable on angiographic restudy in ischemic heart disease patients who had not developed a myocardial infarction (Ambrose et al, 1988).

In each of these conditions one might conclude that compensatory collateral function was adequate. Furthermore, in ischemic heart disease, collaterals enlarge in relation to coronary stenoses. Consequently, other facts must be considered:

1. Apparently healthy people, at their first symptoms of ischemic heart disease, generally show coronary damage marked by one or more chronic critical stenoses or occlusions which had preexisted for months or years. Again, one should state that (a) at least until the first symptoms, the collateral compensatory function was adequate allowing a normal, often stressful, life and (b) in the natural history of ischemic heart disease its clinical onset happens, in general, in the presence of enlarged collaterals because of preexisting severe atherosclerotic lumen reduction. Therefore, in most "acute coronary syndrome" patients, we are dealing not with normal anastomotic channels but with highly enlarged and functioning ones. However, the possibility exists that even normal collaterals may assume an immediate compensatory flow redistribution. An assumption proved in normal subjects who undergo surgical ligation of a lacerated coronary artery following a chest wound yet do not develop a myocardial infarction (Pagenstecher, 1901; Bradbury, 1942; Zerbini, 1943; Bean, 1944; Carleton et al, 1954; Parmley et al, 1958).

Experiments in dogs (Gregg, 1974) confirm postmortem deductions in humans. After an abrupt occlusion of a normal coronary artery, collateral indices such as coronary pressure distal to occlusion and collateral flow into the ischemic area showed a variable, small collateral circulation ranging from 10-32% of normal aortic blood pressure and coronary inflow for minutes to hours. During the first 24 hours following occlusion subendocardial collateral flow increases and often doubles. Gradual coronary stenosis, on the other hand, induces a large increase in collateral indices without evidence of myocardial damage. The release of a coronary occlusion that lasted from 7-10 days caused an immediate decline in this increased collateral function; it reached preocclusion levels within 3 to 24 hours. Reocclusion of the same coronary vessel two months later, rapidly reestablished collateral flow at its previous high value (Khouri et al, 1971; Gregg, 1980). The occlusion of a seven day old critical stenosis did not determine any ischemic or functional change because of a greatly increased collateral flow (Khouri et al, 1968). More recently, in a chronic canine model, by ameroid coronary constrictor, the collateral increase was demonstrated (Mills et al, 2000).

All these facts invite a reconsideration of the role of coronary collaterals in ischemic heart disease and the limitation of angiographic studies done in vivo. The latter: (a) are mainly applied to patients selected by disease; (b) cannot visualize the complexity of the intramural system, including anastomoses and (c) are restricted to a selective injection of one coronary artery. This means that few enlarged intercoronary collaterals are demonstrated while others (homocoronary, extracardiac, intercoronary from the "third" coronary artery present in about half of the human hearts) are not, because of competing nonradiopaque blood flow coming from normal artery(ies). Furthermore, angiography in vivo shares the same criticism as postmortem angiography; the overlapping of injected vessels does not permit correct discrimination between collaterals and parent vessels. On the other hand, the imaging in patients of a retrograde, more or less delayed filling, via intercoronary collaterals of an obstructed main vessel, may not indicate nutrient flow redistribution. The main vessel distal to an obstruction may become an almost excluded channel, the blood flow having different and by-passing routes. In other words, the absence of delayed retrograde filling and/or reduced blood pressure distal to a critical stenosis may not be an incontrovertible sign of "absent or poor" collateral flow as shown by repetitive balloon inflations. Equally, an absent or poor collateral flow may be an effect of extravascular compression of intramural vessels, including collaterals by irreversibly hyperdistended or hypercontracted, i.e., asynergic, myocardium.

It seems justified to assume that angiographic imaging pertains to greatly enlarged collaterals only and the parameter "presence or absence" or "poor collaterals" reported in clinical angiographic studies may have little, if any functional meaning. Thus, the greater extent of myocardial fibrosis found in 68% of transmural biopsies sampled at coronary bypass surgery in the absence of collaterals, versus 29% in presence of good collaterals (Schwarz et al, 1982), becomes questionable. However, some clinical angiographic findings support our postmortem ones. Agreement exists that the largest collaterals are related to multivessel critical stenoses or to a previous myocardial infarct. Similarly, a lack of correlation between the type/extension of myocardial damage in pathologic studies corresponds to a lack of correlation between hemodynamics, number and extension of asynergic segments, metabolic factors and the presence/absence of cineangiographic collaterals (Helfant et al, 1970; Bodenheimer et al, 1977; Cohn et al 1980; Vanovershelde et al, 1993).

We will describe (see below) the secondary disappearance of intramural vessels, including collaterals in the infarct necrotic zone; a disappearance proportional to infarct size. This is a factor rarely considered and difficult, if not impossible, to quantify by clinical angiographic studies. On the other hand, we documented an increasing size of surviving collaterals, likely visible by cineangiography and postmortem angiography, (Spain et al, 1963) in an infarcted region; a finding confirmed by experiment (Reimer et al, 1979) and by the following clinical studies. Recanalization of a coronary artery in acute myocardial infarct (AMI) patients using intracoronary thrombolysis is significantly less frequent (28%) in the presence of a goodly number of collaterals with distal injection of the occluded infarct-related coronary artery than in their absence or with poor collateral numbers (55%) (Araie et al, 1990). In another clinical study, the 33% frequency of collaterals in AMI patients with complete occlusion of the infarct-related artery (11% with subtotal occlusion) increased to 90% at the end-point, 10-14 days from onset of symptoms, in patients with persistent coronary occlusion. In contrast, there was a decrease in collaterals from 38 to 7% in patients with sustained reperfusion (Rentrop et al, 1989). In a third study of AMI patients who had persistent angiographic occlusion, 52% had no evidence of collaterals when studied within six hours, while almost all patients studied later (1 to 45 days) presented collaterals (Schwartz et al, 1984). The "disappearance" of collaterals in patients with sustained reperfusion may only indicate a new flow redistribution following the reopening of a main vessel.

Imaging in vivo is a rapidly evolving technology. Future advances in correlative understanding between functional and structural aspects can be expected. We note the apparent benefits related to the cineangiographic demonstration of collaterals, for example, prevention of postinfarct aneurysm formation; reduction of infarct size; improved ventricular function obtained in several studies (Rigo et al, 1979, etc); the high persistence of myocardial viability within an infarct in the presence of collateral flow (Sabia et al, 1992); an increase of retrograde collateral filling in about 90 seconds after sudden occlusion of a critical stenosis by an angioplastic balloon inflation (Rentrop et al, 1985;1988); left ventricular function and clinical outcome after abrupt coronary closure depending upon the location of the obstruction and degree of collateral flow (Rentrop et al, 1989); patients with a myocardial infarction two days to five weeks earlier who developed asynergic wall motion, revealed that 78% of them had improved ventricular function when angioplasty was successful, while 11% with unsuccessful angioplasty did not improve ventricular function. This improvement correlated with the percentage of infarct bed (> 50%) supplied by collateral flow, assessed by myocardial contrast echocardiography, and was independent of the time between infarction and angioplasty (Sabia et al, 1992), repetitive coronary occlusions e.g., five successive prolonged inflations at angioplasty in patients with chronic angina and an isolated critical (≥ 70%) stenosis of the left anterior descending coronary branch in the presence of normal left ventricular function, induced a progressive adaptation to myocardial ischemia with reduction of symptoms and ischemic ECG changes by recruiting collateral channels. The latter was evaluated by ipsilateral and contralateral injection of contrast medium and hemodynamically by occlusion pressure, suggesting an underlying mechanism of myocardial ischemic preconditioning (Deutsch et al, 1990; Cribier et al, 1992; Sand et al, 2000) even in the absence of demonstrable collaterals (Sakata et al, 1997). This myocardial tolerance depends on collateral recruitment and not on intracoronary adenosine infusion (Billinger et al, 1999), the contractile recovery being independent of increased collateral flow (Barilli et al, 1999). The latter observations indicate that acute mechanical occlusion of a critical stenosis by balloon inflation produces a transient arrest of retrograde collateral flow which is restored in a very short time. Since, experimentally, the increased size of collaterals occurs within a few days of inducing a critical stenosis (Khouri et al, 1968), one can assume that enlarged collaterals preexist before angioplastic occlusion. For some still unknown reason, possibly spasm of the parent vessels, or compression of satellite collaterals of the plaque, following this highly traumatic invasive technique, collaterals disappear immediately after coronary balloon occlusion but return to adequate function with a normal ECG and disappearance of chest pain very rapidly. A similar preconditioning has been documented experimentally in the absence of a preexisting significant stenosis. In chronic, instrumented dogs, an increase of collateral size and function, demonstrated by flow indices, were obtained by repetitive coronary occlusions for 2 minutes every 30 minutes continuously, night and day, for 2-9 days. This structural and functional increase in collaterals prevented or reduced regional myocardial asynergy and reactive hyperemia secondary to transient occlusions lasting 5 to 120 seconds, in the absence of a chronic critical stenosis (Yamamoto et al, 1984).

Finally, the vascularization of a plaque as a possible "satellite" collateral system is apparently neglected. However, this angiohyperplasia forms a local satellite network which may play an important role in bypassing a stenosis (see below). The limitation of coronary plastic cast study is that, despite histologic control of the myocardium, the corrosion of the heart did not permit correct evaluation of the nature of any coronary occlusion, identification of the types of myocardial damage and its extent in relation to its total mass, or structural changes of the atherosclerotic plaque in relation to different clinical patterns of coronary heart disease. This stimulated other investigations.

## Nature, Location and Age of Coronary Occlusion

The subsequent step was to establish nature and structure of coronary occlusions seen in the plastic casts. From the files of the Armed Forces Institute of Pathology, Washington D.C., 208 acute infarct cases, 116 unexpected and 112 expected sudden coronary death cases were selected according to the following criteria:

1. Fatal acute infarct documented clinically and histologically with a reaction ranging from early polymorphonuclear infiltration to reparative healing with remnants of necrotic myocardium.
2. Sudden/unexpected death occurred in less than 30 minutes in apparently healthy subjects with postmortem findings limited to coronary atherosclerosis of any grade and acute or old myocardial necrosis.
3. Sudden/expected coronary death in patients with CHD in their history.

Most of these subjects were on active military duty and had periodic medical examinations. This study (Baroldi, 1965) showed that (a) an acute occlusion was caused by a thrombus; (b) the latter was present in less than 50% of cases; (c) it was, generally, located at the level of a preexisting atherosclerotic functional stenosis (≥ 70% lumen/diameter reduction). The conclusion was that a thrombus forms in a vessel that was obstructed by a chronic process and already bypassed by collaterals as observed by plastic cast studies. The consequent hypothesis was that a thrombus is a secondary ineffectual event, as also shown experimentally (Khouri et al, 1968). Other significant findings of this study were the lack of correlation between the age of the infarct (Mallory et al, 1939) and the age of the coronary thrombus (Irniger, 1963), in about 50% of the cases with an occlusive thrombus (Table 2); the rarity of a ruptured atherosclerotic plaque which always occurred in a severely obstructed vessel and an adventitial-intimal leuko-monocytic inflammation at the plaque level.

From these findings it became clear that "coronary heart disease" was a more complex phenomenon than just obstructing a pipe with the need for a systemic, comparative study in different patterns of this disease and in controls inclusive of noncardiac patients and normal subjects dying from accident.

# Chapter 2

# Comparative Pathologic Study

The objectives were first to compare the different morphologic variables in acute and chronic coronary syndromes to establish the morpho-functional significance of the changes found in coronary vessels and myocardium; second, to evaluate these same variables in a noncardiac population using the same protocol and third, to study other diseases which can be considered "human experiments" since they share alterations which should have a dysfunctional effect similar to that presumed for CHD.

This Chapter is a compulsory preface on materials, methods and definitions for the understanding of data and results which will be reported in the next sections.

## Material and Criteria of Patient Selection and Controls

### *Acute Myocardial Infarction (AMI)*

Two hundred consecutive cases of acute myocardial infarction, 100 at Medical School, University of Milan, Italy (Baroldi et al, 1974) and 100 at the Toronto General Hospital, Canada (Silver et al, 1980) were studied. All had the clinical diagnosis established in a coronary care unit by alterations in ECG tracings and blood enzyme levels including isoenzymes. No patient had another form of heart disease or developed the infarction as a complication of a clinical or surgical procedure and none had coronary vascular surgery, angioplasty or intense resuscitation attempts. All hearts at postmortem showed histologic evidence of infarct necrosis with an associated polymorphonuclear leukocyte infiltration. Thus, patients were selected for study only if they had unequivocal clinical and histological evidence of a myocardial infarction without other diseases and/or iatrogenic damage.

Cases where the AMI was the first sign of CHD were distinghished from those where the AMI was associated with chronic ischemia. A further distinction was made between cases with and without monofocal, extensive (≥ 10 percent of the left ventricular mass) myocardial fibrosis. The latter was considered an acceptable hallmark of previous "silent" AMI. Subjects without either a history of CHD or extensive myocardial fibrosis who had their first symptom in apparently normal conditions were grouped as 1st episode of coronary heart disease and those with extensive myocardial fibrosis with or without history of CHD as 2nd episode or chronic coronary heart disease.

### *Chronic Coronary Heart Disease*

This group comprised 50 patients, all of whom died in Toronto within 25 days of aortocoronary bypass vein graft surgery for clinically documented coronary heart disease with angina pectoris. In all instances death was caused by congestive heart failure.

## *Sudden/Unexpected Coronary Death (SD)*

### Sudden Death Definition

The term sudden includes two concepts: a chronological one, in the sense of a death which occurs in a short interval; the other is precognitive because of a lack of symptoms or signs that might indicate an incipient death; therefore, the death is also unexpected. Already, by 1707, in a report Lancisi clearly defined different types of death:

"Huiusmodi vero absoluta cessatio motuum animalium & abscessio animae a corpore, quanquam cogitatione citius perpetua contingant; nihilominus tamen, vulgaris consuetudinis clariorisque doctrinae gratia, mors distinguitur in naturalem, immaturam, & violentam; singulae vero in lentas, & subitaneas, in praevisas, ac praesensas, denique in improvisas, insensiles, atque inopinatas" ("Indeed this absolutely complete cessation of animal movements and this departure of the soul from the body, even though it happens at all times more swiftly than thought itself, is nevertheless divided for the sake of common parlance and for greater clarity of teaching, into natural, untimely and violent death, and those again individually into slow and sudden death, into those that are foreseen and forefelt and finally into such as are unforeseen, imperceptible and unexpected"; (translation by White et al, 1970).

Lancisi's clear-cut and still up-to-date classification, needs a few comments. First, the meaning of natural versus violent may be ambiguous. Natural is anything that happens in the natural history of being. This is a very broad concept which may comprise any category of death, including one secondary to violence. Perhaps we should distinguish the following types of death:

1. Physiological or genetic death an end result of physiological aging programmed by the genome. We do not know its mechanisms and are often unable to discriminate physiological age changes from those caused by chronic diseases associated with aging.
2. Pathological death due to diseases including malnutrition/starvation.
3. Accidental or violent death due to trauma or any equivalent (wound, poisoning, etc) affecting healthy subjects.
4. Unexplained death when both clinical and postmortem findings are negative or insufficient to explain its cause.

Lancisi's statement: "Non utilis modo, sed maxime necessaria Medicis videtur scientia praecognitionis repentinarum mortium, cum nostrae (a) Praeceptor Artis clare ostendat, eum non solum culpa vacaturum, verum etiam boni Medici nomen, atque admirationem consecuturum, qui, cum omnes sanos facere non possit, futura saltem praesentiat atque praedicat" ("The science of the precognition of sudden deaths is seen to be not merely useful but extremely necessary to physicians, since the teacher of our Art (Hippocrates) clearly shows that man not only absolves himself from all blame, but acquires the name of and the admiration owed to a good physician, when he, unable to make everyone well, at least divines and foretells what is about to happen" (translation by White et al, 1970) is an invitation to establish more precisely the frequency of sudden death in any disease.

We note many definitions of sudden death reported in the literature with the following presenting various points of view:

> "... rapid and unforeseen termination of an acute or chronic disease which has in most cases developed in a latent manner (Brouardel et al, 1902)".
>
> "An individual who died due to natural cause and who was not restricted to his house, hospital or other institutions and who was able to function in the community 24 hours prior of death. The time interval for the onset of the fatal event even until death was less than 24 hours (Kuller et al, 1975).

"Death occurring within one hour, "early" within 24 hours" (Fulton et al, 1969).

"Instantaneous death within 30 seconds, sudden death in minutes to 24 hours" (Friedman et al, 1973).

"We take the colloquial definition "sudden" to mean an unexpected or unusual death which was sudden in general terms and which may or may not have been witnessed, but which poses a mystery for explanation" (James, 1973).

"Witnessed death within one hour of the onset of acute symptoms" (Goldstein, 1982).

" A natural (i.e., nontraumatic) event that is known to have occurred within one hour of the onset of symptoms in a previously healthy person. Use of the term in any other way (to include persons dying , e.g., up to 24 hours after the onset of symptoms) must state the definition explicitly and completely (Hackel et al, 1993).

In outlining a definition of sudden death, the question is whether we really need to establish chronological boundaries. Survival time, i.e., the period between the onset of the terminal episode and death, is considered a discriminating parameter in most of the preceding definitions. Timing the fatal episode is obviously important in evaluating many variables, to help understand the sequence of pathogenic mechanisms and changes secondary to terminal events. However, to include or exclude cases based on this parameter seems unjustified, if not misleading. In the present era of emergency hospital services to allow a 24 hours period before death is too long, because adequate clinical investigation could be carried out in that period. On the other hand, death in a 30 second period can be determined only in very limited circumstances; while one hour or even less may be sufficient for a clinical diagnosis on a patient in hospital and two hours may not be enough to do so for a subject who is out-of-hospital. Accordingly the parameter "survival time" was not included in the criteria of selection (see below) in our sudden coronary death study.

Two basic notions pertain to sudden death. First, its mystery from the clinical standpoint and second, its occurrence in apparently healthy people as well as in those in various phases of a clinically recognized disease. Any study of sudden death should consider this distinction to gain more precise knowledge about the phenomenon. In terms of expectancy, sudden death in a "healthy" athlete during a competition may be quite different from sudden death in a patient with known chronic ischemic heart disease. In other words, a correct approach would distinguish between a first episode and a secondary event in which complications and/or iatrogenic effects may change the natural history of a disease.

On that basis the definition of sudden death which we prefer is a death that is rapid (without any specific chronologic limit) and unexpected or unforeseen - both subjectively and objectively - which occurs without any clinical evaluation, and in apparently healthy people (primary or unexpected or non forseeable sudden death) or in patients during an apparently benign phase in the course of a disease (secondary or expected or forseeable sudden death). Keep in mind that in the present etiologic and pathogenic uncertainty, any definition is only a working one that aids a better selection of material for study. At present, uniquely objective data are postmortem findings and, in a select group, electrocardiographic changes in monitored patients or clinical follow-up in people resuscitated from sudden death.

**Sudden Unexpected Coronary Death without Resuscitation Attempts**

Here death at autopsy was attributed to coronary disease and its complication, so called sudden coronary death. This term is in harmony with the classic pathogenic viewpoint that any coronary arterial obstructive lesion leads to myocardial ischemia with consequent structural and functional damage to the cardiac pump. Already, in 1761 Morgagni, correlated obstructive change of the coronary arteries with chest pain.

In our study (Baroldi et al, 1979) 208 cases of unexpected sudden coronary death (SD) were selected at the Forensic Institute of the Medical School of the University of Milan according to the following criteria:

a. all SD cases were witnessed and occurred outside hospital.
b. a reliable family and personal history was obtained by careful interview of witnesses and family members .
c. subjects were participating in normal and usual activity and were not under medical care or taking drugs for any reason. They had no history of any manifest disease which could be related to death and had not received any medical assistance, therapy or resuscitation attempts during the final episode.
d. significant postmortem findings were absent in all organs other than the heart.
e. the only cardiac lesions demonstrated at autopsy were coronary atherosclerotic plaques with lumen reduction of any degree and/or myocardial necrosis or fibrosis with or without cardiac hypertrophy.
f. in all SD and normal control cases (see below) tests for poisoning or intoxication as a cause of death were negative.

The main reasons why only untreated and apparently healthy subjects were selected for the SD study was first, to avoid any superimposed iatrogenic effect due to therapeutic maneuvers and second, to observe phenomena at their earliest without complications due to other secondary acute or chronic events. The main criticism of this type of selection is that clinical information (family and personal histories) is limited and questionable both because of a nonqualified source and the frequent habit of subjects to minimize or equivocate symptoms. Nevertheless, as for infarct cases, we distinguished SD 1st episode of coronary heart disease vs 2nd episode or chronic cases in absence or presence of an extensive myocardial fibriosis ($\geq$ 10%) respectively.

### Sudden Unexpected Coronary Death versus Resuscitation Attempts

The previous group included sudden/unexpected coronary death without resuscitation attempts to avoid iatrogenic changes. More recently, another 25 cases of sudden/unexpected coronary death were studied in collaboration with Dade County Medical Examiner Department, University of Florida, Miami, USA. They were selected and examined as the previous 208 cases, with the aim of comparing cases without and with resuscitation attempts monitored by electrocardiogram by a rescue team.

## *Other Noncoronary Diseases and Accidental Cases Studied*

To explore the functional meaning of the morphologic variables found in coronary heart disease in general and in sudden coronary death in particular, several coronary and noncoronary conditions needed to be matched with the cases defined above adopting appropriate selective criteria and the same method of examination. The following patterns were studied—and others are under investigation—since several histologic signs seen in sudden coronary death are also found in noncoronary diseases. This matching reported in the appropriate chapters could aid in interpreting cause and pathogenesis of CHD.

### Sudden/Unexpected Death in Silent Chagas' Disease

This group includes 34 apparently normal subjects who died suddenly and had serum positive for Chagas' disease postmortem. Contraction band necrosis and severe myocarditis were the main findings (Baroldi et al, 1997)

**Brain Hemorrhage**

Twenty-seven noncardiac patients with intracranial brain hemorrhage due to rupture of a berry aneurysm. Hemorrhage of the brain activates the adrenergic system with contraction band necrosis (Baroldi et al, 1997).

**Transplanted Heart**

Forty-six patients with orthotopically transplanted hearts with a range of survival from less than 7 days to more than 365 days. Denervation implies an increased sensitivity to catecholamine (Baroldi et al, 2003).

**Acquired Immunodeficiency Sydrome (AIDS)**

This group, formed by 38 cases of AIDS is a model of longlasting hospitalization with opportunistic infectious diseases and emergency therapy (Baroldi et al, 1988).

**Congestive Heart Failure**

The hearts excised from 144 patients (63 coronary heart disease, 63 dilated cardiomyopathy, 18 valvulopathy) with irreversible congestive heart failure who had undergone heart transplantation, formed this group. Surgical excision under anesthesia excludes any agonal effect (Baroldi et al, 1998).

**Normal Population Dead from Accident (AD)**

Ninety-seven cases from the same Forensic Institute of Milan were examined as normal controls (Baroldi et al, 1979)

**Cocaine Abusers/Overdose**

This group comprised twenty-six cases without cardiac disease and without or with minimal coronary atherosclerosis, each of whom had a history of cocaine abuse (Fineschi et al, 1997).

**Carbon Monoxide Intoxication**

These twenty-six people, all without significant coronary atherosclerosis, were found dead at home, (9 suicides). This group was studied to match ischemia versus hypoxia (Fineschi et al, 2000)

**Head Trauma**

Forty-five cases of death out-of-hospital after head trauma in normal subjects all without coronary atherosclerosis. (Baroldi et al, 1997)

**Electrocution**

Twenty-one normal subjects without coronary atherosclerosis who died out of hospital (Baroldi et al, 2001).

We note that none of individuals in the latter five groups were subject to resuscitation attempts.

***Noncardiac Diseases***

These included 100 patients dead of noncardiac diseases (brain infarction/hemorrhage, pneumonia etc) at Milan University Hospital (Baroldi et al, 1967).

## Method of Examining the Heart

In all cases an autopsy was performed between 14 and 74 hours of death, the body being refrigerated at 4° C before examination.

All organs, the aorta and its main branches were carefully examined. The heart was removed from the body, washed, weighed and transverse (left to right margin along the atrioventricular groove) and longitudinal (from aorta to apex) diameters were measured. It was then fixed undistended for 24 hours in 10% buffered formaldehyde solution. After mild fixation coronary arteries and their main branches on the surface of the heart (extramural or subepicardial coronary arteries or branches) were cross-sectioned at 3 mm intervals along their whole course. Samples for histologic examination were taken systematically at the origin of left main coronary artery (LMA), left anterior descending branch (LAD), left circumflex branch (LCX), right coronary artery (RCA) posterior descending branch, (PD), second distal LAD, and at the marginal and middle portion of the posterior tract of the RCA. Furthermore, any 3 mm cross-section showing luminal stenoses with the naked eye were sampled for histology. All samples were placed in 10% buffered formalin to complete fixation and decalcified when needed.

Each heart was cut by a machine into slices 1 cm thick parallel to the posterior atrioventricular groove proceeding from apex to base, with the last section at the upper level of the left ventricular papillary muscles (usually 3-4 cm from the atrioventricular groove). After another 24 hours fixation, sliced hearts were examined and the location, type and size of any myocardial damage recorded. Photographs of the slices, made on a grid divided into 1 cm squares, were enlarged and any areas of acute infarction or fibrosis and the total area of each slice were measured using a polar planimeter. In this manner, the size of an infarct or scar, expressed as a percentage of total left ventricular mass including the whole septum, was calculated. Furthermore, the thickness of the left anterior ventricular wall was measured in the most basal heart slice. Histologic sections were used to establish the edges of an infarct or scar when assessing the affected area.

In each heart the entire ventricular wall at the basal and median levels of the anterior, lateral and posterior walls of both left and right ventricles, the anterior and posterior left and right papillary muscles, the anterior and posterior interventricular septum and the left and right atria were examined histologically. Furthermore, any naked eye lesion in the myocardium was estimated and sampled for histology. An average of 40 sections in 18 different cardiac areas were examined per heart. Both coronary arteries and myocardium were stained with hematoxylin and eosin; and when necessary by Movat pentachrome, Weigert elastic, Mallory and PAS stains.

In sudden death cases the conduction system was excised according to the method of Lev. Systematic samples were taken of the sinus node, the atrio-ventricular node, His bundle and its branches.

The 100 AMI hearts studied at the Toronto General Hospital had a postmortem injection of barium sulphate (Micropaque, Damancy and Co., Slough, UK) into the coronary arteries at a pressure of 120 mm Hg. Radiographic images were made in the anteroposterior and both left and right anterior oblique views.

## Analysis of Extramural Coronary Arteries

### *Physical Variables*

In all histologic sections of coronary arteries the following parameters were evaluated:

#### Intimal and Medial Thickness

The maximal intimal thickness was measured histologically in microns by a micrometer. We distinguished the following types of intimal thickening: 1) a physiologic one; 2) that observed in atherosclerosis and 3) a nonatherosclerotic obstructive intimal thickening. Minimal and maximal thicknesses of the media were established in the same manner, and a ratio "maximal/ minimal medial width x 50" was calculated.

### Lumen Reduction by Atherosclerotic Plaque

The degree of lumen reduction found histologically in a coronary artery was expressed as a percentage and measured as a reduction in luminal diameter. This method was chosen in preference to measuring the cross-sectional area because an atherosclerotic plaque may distend a vessel wall. The rationale is to compare the normal lumen of a vessel with the residual one. The major and minor diameters of the residual lumen were measured in each section of a coronary artery using a micrometer and the results averaged. That average diameter was related to the average luminal diameter obtained in a plastic cast study of coronary arteries from normal hearts (Baroldi et al, 1967).

We are aware that no method of establishing the degree of a coronary artery stenosis is entirely satisfactory. When postmortem injection is not performed a criticism is the lack of fixation of vessels under pressure. We noted no significant difference in the distribution of the degree and number of stenoses found in two series of 100 AMI cases one without (Baroldi et al 1974) and the other with postmortem coronary injection under pressure followed by fixation (Silver et al, 1980) (Table 3). What is needed is a reproducible method that permits comparison between different populations to establish the trend of variable "lumen reduction". In any method of measurement, either in vivo or postmortem, post- or prestenotic dilation must be considered (Rodbard, 1956,1971). A mild stenosis or a normal lumen in a plaque may result in a severe stenosis when calculated by cross sectional area. The concept that plaque enlargement ia a compensatory mechanism (Glagov et al, 1987) is questionable. Most coronary patients show a severe lumen reduction, those with a minor stenosis did not have enlarged plaques.

### Luminal Stenosis

A luminal stenosis in a coronary artery was defined as being severe, functional or critical (capable of reducing flow) when it was equal to, or more than, 70% (a 70 % lumen/diameter stenosis roughly corresponds to a 90 % lumen/area stenosis). The stenosis was mild when lumen reduction was less than 70% lumen/diameter. To compare stenoses in each main vessel among groups with different causes of death the maximal lumen reduction found in a vessel was considered. This allowed us to evaluate any degree of stenosis in one or more main vessels against several parameters (e.g., infarct size, survival, etc).

### Length of the Stenosis

The length of a maximal luminal stenosis in the gross was calculated in millimeters by judging its extension into sequential 3 mm cross-sections.

### Type of Stenosis

Histologically, a luminal stenosis was defined as concentric when the residual lumen was centrally located or when it was lateral but still encircled by pathologic tissue, or semilunar , when part of the arterial wall was normal.

## *Luminal Thrombosis*

In its early stage a thrombus (acute thrombus) is mainly composed of platelet aggregates, fibrin, and some polymorphonuclear leukocytes; later, in healing, it shows different stages of organization with eventual luminal fibrosis and recanalization (old thrombus).

**Occlusive Thrombus**

We included in this category any thrombus that "completely" occluded a coronary artery lumen. The morphology of a thrombus may change with location in the lumen being completely occlusive in one section and partially occlusive (75%) in another (e.g., in its "tail"). Because of this we included partially occlusive with completely occlusive thrombi.

**Mural Thrombus**

Such thrombi occluded less than 50% of the lumen. In general, acute mural thrombi were formed by thin, mainly fibrinous, lamina which did not reduce the lumen significantly. Old mural thrombi showed different stages of organization.

### *Morphologic Variables*

**Type of atherosclerotic plaque.** A plaque was defined as atheromatous or fibrous according to whether atheromatous material or fibrous tissue (without basophilia and/or atheroma) predominated in it. The following variables were also considered:

**Proteoglycan accumulation** presented as a pale, amorphous basophilic substance without cellular reaction. It is found in the external layer of the thickened intima deep to a fibrous cap.

**Atheroma,** consisted of a combination of lipoprotein material, foam cells and crystalline, cholesterol clefts.

**Calcification** appeared as basophilic granules of various size or as a plaque of darkly basophilic material replacing intimal tissue.

**Intimal hemorrhage** was indicated by extravascular red blood cells found in lesions in various amounts.

All of these variables were defined as mild when, in total, an individual one involved one-quarter or less of the circumference of the vessel wall, moderate if half was involved and severe if more than half of the circumference was affected.

**Intimal vascularization** was characterized by finding capillary-like vessels of varying diameter in the thickened intima. According to their number, vascularization was considered mild if less than three lumina were seen, moderate if four to six were present or extensive if more than six.

**Intimal and adventitial lymphocytic infiltration** was marked by inflammatory cells, mainly small lymphocytes and plasma cells in the intima and/or adventitia. An inflammatory reaction was considered mild when only a few, scattered cellular elements were found, moderate when few but well profiled foci of lymphocytes were present, and severe when a massive inflammatory reaction was seen. When plasma cells and lymphocytes were located around nerves adjacent to the tunica media, we defined it as medial neuritis.

## Analysis of Intramural Arterial Vessels

In each histologic myocardial section the status of intramural (or intramyocardial) arterial branches of any type, including the terminal bed and veins, was investigated. The following main changes were considered:

### *Thromboemboli*

**Platelet aggregates** in a vessel lumen consist of faintly basophilic granular material, formed by very small, roundish elements with different degrees of aggregation often in dissolution. No demonstrable fibrin was associated with them. To calculate their frequency 16 histologic sections of myocardium (5 left ventricle, 5 right ventricle, 4 interventricular septum and one for each atrium) were selected at random from all histologic slides of each case. They included the sinus node in 120 SD, and 63 AD cases and the AV-node bundle of His in 180 and 95 cases respectively. In each section the number of arterial intramural vessels partially (≥70 percent of the lumen) or completely occluded by platelet aggregates were counted by screening the entire section at 250 x magnification. A total of 3328 sections in 208 sudden/unexpected coronary death, and 1552 in 97 normal individuals dead from accident were examined (Baroldi et al 1980). The presence or absence of venous platelet aggregates and blood stasis were also estimated.

**Fibrin-platelet thrombi or emboli** presented an association, in variable proportions, of fibrin and platelets. They may form in situ (thrombi) or have origin from a proximal source (emboli).

### *Vascular Stasis*

Due to postmortem changes and technical artifacts it is difficult to objectively quantify the amount of blood in myocardial tissue. To have a rough estimate of stasis at death, intramural stasis, was defined as arterial or venous or both when in each histological section at least five intramural arterial or/and venous vessels respectively were well filled by red blood cells.

### *Medial Hyperplasia Obliterans*

By this definition we indicate a medial change affecting small intramyocardial vessels. It consists of a hyperplastic process with development of longitudinal bundles of smooth muscle cells found mainly in the outer media that cause luminal stenosis. We consider the finding of fibrous tissue penetrating into and replacing medial muscular tissue a late stage of the process. In this condition the intima and internal elastic membrane are usually normal and only occasionally are fibrous intimal thickening and degenerative alteration of the elastic membrane seen. This pattern was defined as minimal when only one vessel showing these changes was present in at least one of the 18 areas examined, moderate when two vessels were affected and severe when more than two vessels were seen (Baroldi, 1986).

## Analysis of Myocardial Changes

The different forms of myocardial necrosis observed in coronary heart disease and stages in their healing were evaluated as follows:

### *Acute Myocardial Necrosis*

Infarct necrosis was grossly estimated in percent of the left ventricular mass (see above). Coagulative myocytolysis, or Zenker necrosis or contraction band necrosis (CBN) and colliquative myocytolysis (or myocytolysis) were judged minimal when less than five foci were observed in one histological myocardial section, moderate when a similar number of foci were seen in two or three sections and extensive when that number was present in four or more sections. In the more recent studies CBN was calculated as the number of foci and myocells affected x 100 $mm^2$ in each histological section. Colliquative myocytolysis was divided into the following grades: 0, absent; 1, occasional or small groups of myocells affected. 2, less than and 3, more than 50% of myocells involved with total or subtotal disappearance of myofibrils in the internal half of each histologic myocardial section.

### *Myocardial Fibrosis*

Myocardial fibrosis was classified as recent by the presence of fibroblasts and vessels or old when it was dense, hypo- or acellular and avascular. It was estimated as minimal when only a few foci were detected histologically, moderate when its extension was less than 10% of the left ventricular mass and extensive when more than 10%. Furthermore, more recently, it was calculated in percentage of the each 100x fields of the total histological area by an orthogonally bisected ocular. Since myocardial scar may transform into adipose tissue (Baroldi et al, 1997) this is the only reliable method to measure the extent of myocardial fibrosis histologically. For comparative purpose amongst all examined conditions, a ratio between total fibrotic area/total histological area in $mm^2$ x 100 was assessed. This fibrous index determines in percent the amount of histologically viable myocardium versus scar. In absence of reliable clinical history, a monofocal scar greater than 10% of the left ventricular mass was considered an acceptable sign of an old infarct for a distinction amongst 1st episode and chronic cases of acute coronary syndromes.

### *Cardiac Hypertrophy*

In our material heart weight was reported in "100 g" classes (< 200, 200-299,300-399 etc). However for comparative purpose and to avoid indices related to body-weight and/or height, we adopted Linzbach's distinction between physiologic (< 500 g) and pathologic (≥500 g) hypertrophy. In general a heart exceeding 500 g has a pathologically increased mass. A heart was defined as atrophic when its weight was less than 250 g for men and 200 g for women (normal average weight 300 and 250 g respectively (Silver MM, 1991).

### *Myofiber Disarray*

Architectural disarray of myocardial cells is the typical change seen in hypertrophic cardiomyopathy (Teare, 1958). Its presence was considered pathological when its extent was higher than 20% of the histological area.

## Statistical Analysis

All variables and their ratings were recorded on original cards. The data were processed by an IBM 370/168 computer. Analyses were accomplished by nonparametric tests (Bishop et al, 1980). The significance of first and superior-order associations were investigated by log-linear model. When a "fit" of specific models was obtained, further analysis on the pertinent contingency tables was done by residual and lambda parameter analyses. For subject analysis, chi square tests and "filling" to binomial distribution function were used. Possible associations among morphological variables were tested, using two codes for comparison. They compared no change versus mild + moderate + severe changes (sensitive code), and no change + mild change + moderate change vs severe change (specific code). To avoid a tedious repetition of chi-square values in the text, a significant result indicates one where the P value is < 0.05.

CHAPTER 3

# Natural History of the Human Coronary Atherosclerotic Plaque and Related Forms of Myocardial Injuries

The first need was to reconstruct the natural history of the morphology of the coronary atherosclerotic plaque seen in humans with and without CHD. Since different arterial vessels have dissimilar hemodynamics, wall structure and nervous control, each artery has to be studied independently because the course of the atherosclerotic process could vary, in the aorta, cerebral arteries and coronary arteries. For instance, in a comparative study of 40 hearts with an acute myocardial infarct without cerebral diseases versus 41 cases of brain infarct/hemorrhage without cardiac diseases, the frequency of a severe stenosis (≥70%) was 88% for coronary arteries and 10% in brain arteries in the first group and 53% and 14% respectively in the second group. Note (a) the very high frequency of severe coronary stenosis in the brain patients without heart disease and (b) the very low frequency of severe obstructions in the cerebral arteries of both groups (Fig. 5; Antoci et al, 1980).

## Coronary Atherosclerosis

Blood vessels are very sensitive structures which respond to hemodynamic changes. Thus endothelial nuclear shape and orientation e.g., elongated, flow-direction oriented nuclei in segments with stable flow; round, less ordered nuclei in segments with unsteady, turbulent flow and possibly the density of endothelial nuclei depend upon stresses secondary to flow dynamics (Flaherty et al, 1972). In general, the architecture of a vessel wall is proportionate to the latter (Burton, 1954) and may change according to the nature of variations in flow dynamics (Rodbard, 1971). Types of stresses which act on the vessel wall are compressional with a radial direction, tensile with circumferential and longitudinal directions; and shearing, which depends on flow velocity and viscosity and is caused by the drag of flowing blood acting parallel to the vascular surface (Fry, 1969)). When stresses reach a critical point, structural changes in the vessel wall can be expected (Langille, 1991). In particular, an increase in shear stress stimulates vasodilation in normal coronary arteries, limiting this stress at the endothelial surface. This is in contrast to atherosclerotic arteries in which vasodilation is reduced and major shear stress is likely (Vita et al, 1989).

Human coronary arteries may be divided into two functionally divergent systems in relation to cardiac and flow dynamics; the extra- and intramural vessels. They appear to be a good model in which to study the relationship of flow dynamics and atherogenesis. In extramural arteries a distinction must be drawn between a) physiological intimal thickening; b) nonatherosclerotic obliterative intimal thickening; and c) segmental atherosclerotic obstructive intimal thickening based on clear-cut structural differences induced by each of these types of intimal processes.

*The Etiopathogenesis of Coronary Heart Disease: A Heretical Theory Based on Morphology, 2nd ed,* by Giorgio Baroldi and Malcolm D. Silver. ©2004 Eurekah.com.

## *Physiologic Intimal Thickening*

Normally, the intima of extramural coronary arteries shows a histologic pattern that is not seen, or at least not is as well developed, in other muscular arteries. Although the role of this change in the pathogenesis of atherosclerosis is controversial, the finding must be distinguished from changes seen in atherosclerotic plaques (Silver et al, 1980; Angelini et al, 1990; Baroldi, 1991). It consists of an intimal thickening starting at birth (Bork, 1926; Dock, 1946) and undergoing progressive changes with age. According to Wolkoff (1929), two layers more or less divided by an elastic lamina are readily recognizable in the first decade of life. The outer one, defined as elastic-muscular is formed by the splitting of the internal elastic membrane and by a proliferation of medial smooth muscle fibers through fenestrations in the latter. The proliferating smooth muscle fibers assume a generally longitudinal disposition. The inner or elastic hyperplastic layer is formed by elastic fibers derived from the separating lamina. It must be noted that in the first decade of life, such intimal thickening frequently does not exceed medial width. It is more prominent at branching sites, does not generally involve the whole vessel circumference, and shows a great variability in different subjects and in different segments of the same vessel, being more prominent in men than women (Dock, 1946; Fangman et al, 1947; Moon, 1957; Schornagel, 1956). In particular, Dock (1946) found a three-fold greater frequency of this intimal thickening in male newborns, a finding not confirmed by Minkowski (1947) who observed a greater frequency in males only in subjects older than one month. In subsequent decades of life, a thickening of the elastic hyperplastic layer occurs with the appearance of a subendothelial fibrous layer which becomes prominent after the fourth decade. Intimal thickening at this time may exceed medial thickness (Wolkoff, 1929; French et al, 1962; Geer et al, 1968; Likar et al, 1969; Vlodaver et al, 1968).

Three stages in the evolution of this physiological intimal thickening can be recognized: (1) an early stage characterized by nodular proliferation in the intima of medial myocytes and elastic lamellar splitting; (2) a stable, hyperfunctioning stage, mainly characterized by diffuse myocellular and fibro-elastic hyperplasia; (3) an exhaustion stage, in which the intima becomes fibrotic (Fig. 5). In this proliferative response, the predominant role of medial smooth muscle cells has been interpreted as that of a multifunctional medial mesenchymal cell capable of contraction, proliferation, migration, colonization, and synthesis of collagen, elastin, ground substance and basement membrane material (Wissler, 1967).

Some authors believed physiological fibro-elastic-muscular intimal thickening represents early atherosclerotic damage (Ehrich et al, 1931; Fangman et al, 1947; Moon, 1957) or, is secondary to (1) an inflammatory-allergic processes (Minkowski, 1947), (2) platelet micro-thrombi deposition (Likar et al, 1960), or (3) hemodynamic stress (Spalteholz et al, 1931; Schornagel, 1956; Vlodaver et al, 1967; Baroldi, 1981). However, such circumferential intimal thickening, normally found in adults, is better considered a component of postnatal vasogenesis (Vlodaver et al, 1967) related to the peculiar flow dynamics in extramural coronary arteries, i.e., systolic filling without or with minor intramural discharge due to myocardial systolic contraction which increases all types of wall stress. It is not necessarily associated with true atherosclerotic change. Variations in the degree of intimal thickening in different subjects most likely depend upon individual variations in flow dynamics. Their importance is confirmed by the absence of such intimal thickening in the intramural arterial system and in those tracts of extramural coronary arteries covered by myocardial bridges ("Mural coronary artery" after Geiringer, 1957a). One presumes that systolic contraction of the latter counteracts the action of dynamic stresses on the arterial wall, dampening tridimensional expansion and hindering any proliferative response. However one can not exclude that in this response a different intensity of neural, possibly adrenergic control on arterial wall tone may play a determining role. This may explain divergencies amongst individuals, different ethnic groups (Vlodaver et al, 1969) and species (French, 1962; Geer, 1968) with a similar distribution of extramural coronary arteries; for instance we have not observed the process in dogs (Fig. 5). Increased intimal thickness in angiographically normal coronary arteries has been demonstrated by intravascular ultrasound imaging in patients with spastic angina in the absence of any traditional risk factors (Miyao et al, 2000).

### *Nonatheroscerotic Obliterative Intimal Thickening*

Nonatherosclerotic obliterative intimal thickening is a diffuse pathologic process of unknown cause which affects the whole intima of extramural coronary arteries. It may produce extremely severe concentric vascular stenosis and can be observed in conditions, such as (a) coarctation of the aorta (Vlodaver et al, 1968); (b) in transplanted human hearts (Thomson, 1969; Cooley et al, 1969; Bieber et al, 1970; Smith et al, 1987; Billingham, 1988; Baroldi, 1991; Rose et al, 1991; Pethig et al, 1999); (c) in transplanted dog hearts (Kosek et al, 1969), and (d) in aorto-coronary saphenous vein grafts (Johnson et al, 1970; Marti et al, 1971; Vlodaver, 1971; Brody et al, 1972; Kern et al, 1972; Virmani et al, 1991). We note that similar intimal/ medial changes affect the vessels of other organs in a variety of both human and experimental conditions such as hypertension (Spiro et al, 1965; Oka et al, 1967; Esterly et al, 1968; Still, 1968; Constantinides, 1970; Huttner et al, 1970; Wolinski, 1972), following catecholamine administration (Szakacs et al, 1959), by varying flow volume (Rodbard, 1956; Hassler, 1970; Schaper et al, 1972); following trauma (Hassler, 1970); and in specific degenerative (e.g., juvenile intimal sclerosis) or infectious-immune diseases (e.g., rheumatic fever).

This obliterative intimal thickening has been interpreted as a variant of atherosclerosis. Some investigators, wrongly in our opinion (see below), speak of "accelerated atherosclerosis" affecting vein grafts (Bulkley et al, 1977) and the coronary arteries in transplanted hearts (Ip et al, 1990; Rose et al 1993). Long-term survivors of cardiac transplantation (Graham et al, 1972; Rider et al, 1972), and experimental cardiac allografts where the recipient animals are fed a cholesterol rich diet (Alonzo et al, 1970) do develop atheromatous deposits in this intimal thickening. However, the latter is likely a late, secondary process. "Acceleration" of a process means that its history occurs in a shorter period than usual. However, all components of the process must be present. If not, the process is a distinct entity, with its own history. One notes that nonatherosclerotic obliterative intimal thickening often occurs in denervated vessels (transplanted heart, vein graft) suggesting a possible role for the loss of neurogenic control on vessel wall tone. In relating this change to an effect of cyclosporine one should recall that this obliterative intimal process was observed in the precyclosporine era of heart transplantation (Thomson, 1969; Cooley et al, 1969). Nevertheless, both physiologic and idiopathic obliterative intimal thickening, particularly in subjects at risk, may predispose a vessel to atherosclerosis. As dynamic factors appear important in the pathogenesis of atherosclerosis, it is not surprising that intimal thickening and atherosclerosis may have a similar location. However, idiopathic obliterative intimal thickening, even in its last subocclusive stage, is a result of proliferation of smooth muscle cells with minimal or absent elastic hyperplasia, increased ground substance (proteoglycans) and interstitial fibrosis. The internal elastic membrane is intact and not one of the morphologic variables, i.e., hemorrhage, vascularization, atheroma, calcification, lympho-plasmacellular inflammation of the atherosclerotic plaque is obvious.

### *Atherosclerotic Intimal Thickening*

Atherosclerotic intimal thickening (Figs. 6, 7) is mainly a segmental lesion with complications which may involve the media and adventitia (Strong et al, 1968). The frequency and extent of the previously defined morphologic variables of the coronary atherosclerotic plaque were calculated in 3,640 coronary sections sampled from 100 AMI, 50 chronic CHD, 208 SD patients and 97 normal subjects dying from accident (Baroldi et al, 1988). Our findings in normal subjects seem to indicate that coronary atherosclerosis evolves in adult age with minimal or further progression in the elderly (Table 4). Amongst 1,519 sections that did not show any lumen reduction and had an intimal thickness less than 300 μm (physiological thickening) morphologic variables were insignificant (Table 5). More importantly, no subendothelial fatty streaks, lipoprotein/cholesterol subendothelial infiltration, foam cells, mural thrombi, luminal platelet aggregates, hemorrhage, advential-intimal inflammatory infiltrates were observed in any of these sections nor in 1,319 sections where the lumen reduction was less than 70% nor in 743 sections with intimal thickness less 1,000 μm. Only occasionally, subendothelial lipoprotein/cholesterol plus "foam cells" were seen in severe stenoses; as did mural thrombi.

Verifying the trend of morphological changes in relation to intimal thickening and lumen reduction in ischemic and "clinically" normal subjects, a history of the morphology of coronary atherosclerotic plaque was constructed. From the significant associations of first (two variables) and second (three variables) order of variables and the highest chi-square values obtained according to sensitive and specific code (see statistical method), it was possible to outline a tridimensional, i.e., radial, circumferential, longitudinal progression of the atherosclerotic plaque in the general population including ischemic heart disease patients and healthy controls. It is as follows: initially a plaque is a nodular fibrous intimal thickening likely due to smooth muscle cell and elastic fiber hyperplasia with subsequent fibrous tissue replacement. This early fibrous plaque is the only pattern occasionally seen in young people less than 20 years old (Angelini et al, 1990). The second stage is proteoglycan accumulation deep to the fibrous cap. Both fibrosis and proteoglycans are recurrent phenomena being the two basic elements in plaque progression. Subsequently, foam cells and cholesterol clefts and/or calcification appear in the proteoglycan pool, in keeping with the chemical affinity of glycosaminoglycans for lipoproteins and calcium salts (Wight et al 1983). Therefore, a proteoglycan pool may evolve either into a calcified area and/or atheroma. The final plaque pattern is a result of the extension in three directions of these repetitive phenomena (Figs. 6, 7) plus further complications including hemorrhage, thrombosis, etc. A pattern (Morgan, 1956; McGill et al, 1968; Velican et al, 1989) different to that seen in experimental plaques obtained by a hypercholesterol diet or in familial hypercholesterolemia (Fig. 7F).

The genesis, evolution and role of a coronary atherosclerotic plaque in coronary heart disease has been synthesized by Fuster et al (1992) in the following manner:

"The initiation of atherosclerosis may result from blood flow oscillatory shear stress in certain vascular sites (bending points, bifurcations, etc) producing chronic minimal injury resulting in functional alteration of the arterial endothelium type I injury: experimentally, this is potentiated by atherogenic risk factors such as hypercholesterolemia, hypertension, immunocomplexes, viral infections, and tobacco smoke. Such minimal injury leads to accumulation of lipid and monocytes (macrophages), and subsequently, toxic products released by the macrophages produce damage of the intimal surface with denuding endothelium type II injury or damage, which attracts platelets; all of these cells release growth factors, prompting migration and proliferation of smooth muscle cells and producing a "fibro-intimal lesion" or the outside of the capsule of a predominant "lipid lesion." The lipid lesions surrounded by a thin capsule tend to be small and rupture easily, causing type III injury or damage; that is, they are soft and weak, contain large numbers of macrophages, which may release collagenase and elastase to form abscesses, and by their location, are under the effect of flow shear forces. After plaque disruption there is thrombus formation: when thrombi are small, they can become organized and contribute to the growth of the atherosclerotic plaque; when thrombi are large and occlusive, they lead to the acute coronary syndromes. New data suggest that, at the time of plaque disruption, certain "thrombogenic" risk factors modulate the degree of thrombogenicity and, thereby, the growth of the plaque versus the various acute coronary syndromes. Aside from the need for better understanding of the basic biology of atherogenesis, emphasis on identifying and modifying the primary atherogenic and thrombogenic risk factors should continue for primary prevention. Also, new approaches should focus on the identification, stabilization, and regression of the small "lipid plaques" prone to rupture (these are not necessarily angiographically apparent), as well as on the use of better and safer antithrombotic agents for prevention of progression".

This quotation has the merit of synthesizing the current view point (Libby, 1995; Newby et al, 1999; Libby, 2000) on the cause and pathogenic mechanisms of acute coronary syndromes. Our impression, however, is that this model is biased by experimental data obtained in animals fed a cholesterol diet (Ross, 1993). A model which may correspond to the situation in human familial or acquired hypercholesterolemia, but not to findings observed in the general population with or without ischemic heart disease. In the latter, our findings of an absence of platelet aggregates or platelet-fibrin thrombi or subendothelial lipoprotein/macrophage or monocytes infiltration (foam cells) or fissuring in 1519 coronary sections with normal lumen and in 1315 coronary sections with mild (≤69%) luminal stenosis and minor (<600 μm) intimal thickness supports the existence of two types of atherosclerotic plaque each with a different natural history: they are the *hypercholesterol plaque* and the *smooth myocell hyperplastic plaque.* We have already described a possible progression of the atherosclerotic process in the latter. Keep in mind that the nature (monoclonal? Benditt, 1974) and stimulus of the first intimal changes, nodular smooth muscle cell hyperplasia and sequelae, are still not explained just as there is no demonstration of rupture and thrombosis of small lipid plaques. The concept of different types of plaque growth suggests that regression may be different in each type of plaque; a hypercholesterol plaque being more prone to stop or regress if hypercholesterolemia is normalized.

Nevertheless, several points in this history are still obscure. For example, the factor(s) which promote smooth cell hyperplasia (platelet growth factors; Ross, 1979; catecholamines, etc, Velican et al, 1989); the nature of cells participating in plaque growth, i.e., whether endothelial, smooth muscle, monocyte/macrophages, histiocytes or a unique mesenchymal multifunctional cell capable of transforming structure and endocrine activity according to functional need and the role of hemodynamic stresses all require further clarification. An impressive example of the importance of hemodynamic stresses is provided by adult patients with an anomalous origin of a coronary artery from the pulmonary artery with low flow pressure and pulsation. In this condition, the anomalous artery is free of atherosclerosis whereas the one arising from the aorta, may be severely atherosclerotic (Kaunitz, 1947; Burch et al, 1962; Blake et al, 1964). These patients have an unique genetic and enviromental background. Flow dynamics in the two coronary arteries seem to provide the only difference that explains these findings. Again, diphasic flow dynamics may stimulate the neural control of the vessel which may be responsible for the morphologic changes.

An answered question relates to the variability in location of atherosclerotic plaques in the coronary system, despite some preferential sites of formation. In other words, how do etiologic factors act preferentially in vascular segments of one subject while in others different segments are involved, despite an apparently similar wall structure and hemodynamics?

The effects of intimal vascularization on plaque formation are also unknown. In our study, which was in partial agreement with that of Geiringer (1951), vascularization was present in only 24% of sections with an intimal thickness between 300 and 599 μm. Its frequency was maximal (75%) with a thickness between 600 and 1,999 μm, and was a little less with greater intimal thickness. However vascularization was mild in the majority of sections (61%). Our data suggest that it follows plaque formation. Furthermore, in our experience, the possibility that neovascularization is the result of the organization of mural thrombi (Morgan, 1956) can be considered only when found in severe stenoses (for serial section findings see below).

## Inflammation

Another variable of the atherosclerotic plaque is the inflammatory reaction. Its role in atherogenesis must be clarified. A current theory, arising from study of the experimental plaque which develops in the hypercholesterolemic animal, states that atherosclerosis "results from an excessive inflammatory fibroproliferative response to various forms of insult to the endothelium and smooth muscle of (the) artery wall ...... The earliest recognizable lesion is the so called "fatty streak", an aggregation of lipid-rich macrophages and T lymphocytes whithin the inner most layer of the artery wall, the intima ....... The macrophage is the principal inflammatory mediator of cells, acting not only as antigen-presenting cells to T lymphocytes but also a scavenger cell to remove noxious materials and as (a) source of growth regulatory molecules and cytokines." (Ross, 1993) In contrast are the lympho-plasmacellular infiltrates we observed in human plaques from the general population and ischemic patients. They appear to be a distinct inflammatory response to a still unknown agent (Fig 7). Long described in the literature (Morgan, 1956; Velican et al, 1989), these infiltrates have been considered a minor complication related to plaque size (Schwartz et al, 1962) and possibly to reduction of arterial lumen or interpreted an autoimmune process (plasma cell cytoplasm stained with IgG and IgM antisera; Parums et al, 1981, Wal et al, 1989). In our experience this inflammatory infiltration begins to form in the "proteoglycan accumulation" stage of the plaque, being visible in 32% and 62% respectively of mild (< 50%) and moderate (50-69%) luminal stenoses and with an intimal thickening between 300-599 μm. The importance of this lesion is its significantly higher frequency, extension and strategic location around adventitial nerves adjacent to the media in patients with coronary heart disease than in healthy people with an equal degree of coronary stenosis and intimal thickening. Finally, the prevalence of short (≤3 mm) severe stenoses in AMI patients may indicate that this inflammation always found in these subjects, may influence the radial progression of plaque (Baroldi, 1985; Baroldi et al, 1988; Cliff et al,1988).

## Intimal Hemorrhage

Intimal hemorrhage is a real event in the natural history of a coronary atherosclerotic plaque (Fig. 7) and has been considered the possible source of lipoprotein material, vasoactive substance and thrombogenic factors (Paterson, 1938; Morgan, 1956; Velican et al, 1989) or cause of occlusion (Wartman, 1938). Intimal vascularization is a potential source of intimal hemorrhage although (Davies et al, 1984) provided this contrary opinion: "... we have avoided the term "plaque hemorrhage" since it is a source of confusion. "Plaque fissuring" is the term applied to the formation of an opening from the lumen into the intima; it leads to what was known originally as "dissecting hemorrhage" but is actually an intraintimal thrombus not just red cells but mainly fibrin and platelet."

Intimal hemorrhage was observed in 21% of cases with unstable angina, 19% of sudden death and 63% of acute infarct cases in one study (Kragel, 1991). In ours it was the variable with the lowest total frequency (14%) and with the lowest frequency at any level of lumen reduction or intimal thickness. In 289 sections with intimal hemorrhage, 48% were in an infarct-related artery.

## Plaque Rupture

Plaque rupture is another parameter to be considered in natural history. It occurs in atheromatous plaques ("rupture-prone plaques") and is generated by tiny fissures at the periphery of the fibrous cap that covers the plaque's lipid-rich core (Falk, 1992). At this location the plaque is thinner and infiltrated by macrophages. Intimal "macrophage" inflammation has been proposed as a possible mechanism of plaque fissuring (Buja et al, 1994; Moreno et al, 1994; Wal et al, 1989, 1994). Plaque rupture may result in occlusion of the lumen by releasing pultaceous material (already reported by Branwood, 1956) or be associated with overlying thrombus formation (Osborn, 1963; Friedman et al, 1966; Constantinides, 1970; Ridolfi et al, 1977). In our study we did not serially section all examined plaques. Therefore, we have no exact figures about the frequency of plaque rupture. However, coronary occlusion by pultaceous material alone was an exceptional finding and always associated with hemorrhage or thrombus at a different plaque level. A thrombus with subintimal expansion through a break in the intima was also a rare findings.

Plaque fissuring was reported in 89% of 115 coronary vessels with an associated mural or occlusive thrombus in one study (Davies et al, 1984) and in 81% of 25 vessels in another (Falk, 1985). In yet other reports plaque rupture per se was observed in 36% of cases with unstable angina, 19% with sudden death and 75% with acute myocardial infarct (Kragel et al, 1991); or was absent in cases with unstable and stable angina and present in 7% of cases with acute myocardial infarction, 4% with sudden death, 12% with congestive heart failure and 7% of control cases (Arbustini et al, 1991). Keep in mind that by coronary angioscopy 60-80% of patients with unstable angina have complicated atheromata, i.e., rupture, ulceration, thrombus formation (Sherman et al, 1986; Forrester et al, 1987; Hombach et al, 1988).

**Plaque Calcification**

This variable can be easily detected in vivo and erroneously interpreted as a sign of severely obstructive coronary atherosclerosis. Amongst 990 coronary sections with minor lumen stenosis 33% had massive calcification. This lack of correlation between lumen reduction and calcification has been documented in vivo (Sangiorgi et al, 1995), without any prognostic value (Detrano et al, 1999)

**Medial Changes**

In relation to coronary artery spasm, the type, frequency and extension of tunica media damage at the site of an atherosclerotic plaque are all important factors. This point has, generally, been poorly investigated. In our experience, medial changes were seen only at the plaque site and consisted in thickness reduction in concentric plaques. Occasionally, focal distruction associated with inflammatory reaction was observed. In another study a reduction of 70% of the medial area was calculated at the plaque level. (Arbustini, 1991).

**Plaque Regression**

A concept pertinent to the natural history of the atherosclerotic plaque is its regression. This has not been studied specifically in our investigations. A reduction or disappearance of luminal stenosis has been reported occasionally in coronary angiographic studies (Bemis et al, 1973; Gensini et al, 1972; Laks et al, 1979; Rafflenbeul et al, 1979; Haft et al, 1993) and interpreted as recanalization of a thrombus, lysis of an embolus (O' Really et al, 1974) or resolution of vasocostriction or spasm. However, regression of angiographic lesions and reduction of clinical events, i.e., death, infarct, worsening symptoms, were obtained by intensive lipid-lowering therapy in patients with high levels of apolipoprotein B, documented ischemic heart disease and a family history of vascular disease (Brown et al, 1990). Furthermore, the reduction of an experimental lesion after suspension of an atherogenic diet (Wissler, 1978) and the practical absence of advanced atherosclerotic plaques in cachectic people raises the possibility of plaque regression (see above).

In the International Nifedipine Trial on Antiatherosclerotic Therapy, regression marked by decrease in percent diameter stenosis ≥20% was observed in only 4% of 1063 coronary segments when 348 patients with moderately advanced coronary atherosclerosis, i.e., one or few coronary stenoses or occlusion in only one major vessel, were studied by quantitative coronary angiography performed 3 years apart (Jost et al, 1993) . In this study progression of coronary obstruction occurred in coronary segments greater than 2 mm in diameter in a proximal or midartery position and in the right coronary artery. As mentioned previously regression of a plaque may depend from its morphologic type (myohyperplastic vs hypercholesterol).

## Different Forms of Myocardial Injury Related to Coronary Atherosclerosis and Contractile Function

The relationship between coronary atherosclerotic obstruction and myocardium is, generally, referred to as ischemic asynergy, or loss of contraction in vivo and ischemic necrosis postmortem. Our studies enabled us to distinghish three different forms of myocardial cell injury which we related to the contraction/relaxation cycle. They will be described after a brief synthesis of the myocell's functional anatomy.

### *Functional Anatomy of Myocardial Cell*

Myocardial cells are the morpho-functional units of heart muscle. Grossly they form four main muscle bundles, two with internal circumferential configuration and two that are external and helicoidal. Each bundle is anchored to the fibrous framework of the heart located around valves and large vessels. This architectural structure allows twisting-shortening contraction of the myocardium with optimal blood ejection. The heart muscle is not an anatomical syncytium because at its extremities each myocell, a cylindrical structure 50 to 100 μm long and 10 to 20 μm wide, is separated by, and connected with, adjacent myocells by intercalated discs. Nevertheless, if individual cells are not a syncytium anatomically, heart muscle can at least be defined as a functional syncytium capable of rhythmic contraction-relaxation cycles. The function of the normal cardiac pump is achieved by contraction of all myocells coordinated by neurogenic impulses via the conduction system, regulated by intramyocardial nervous reflexes. It begins at the third week of fetal life and when it stops, life ends. Cessation of contractile function is frequently the cause of death.

The contractile apparatus of any single myocell is formed by a bundle of cylindric myofibrils each subdivided in 20-50 subunits (sarcomeres) separated by thin Z lines. A sarcomere constitutes the basic morpho-functional unit and is constructed for rhythmic contraction-relaxation cycles. It is formed by two centrally separated sets of thin or actin filaments (1-μm long) implanted on two limiting Z lines. Parallel to and between the thin filaments are thick L-meromyosin filaments (1.5 μm long) located in the central part of the sarcomere. The thick filaments are not attached to Z lines. Thin and thick filaments have lateral, corresponding digitations, tropomyosin-troponin and H-meromyosin respectively, which are the active sites of the biochemical hinge which regulates contraction-relaxation. This is achieved by a back-and-forth movement of thin filaments which penetrate the other half of the sarcomere by "sliding" on the thick filaments (sliding theory). In diastole the tropomyosin-troponin complex inhibits contraction. The latter is reestablished by $Ca^{++}$ binding to troponin. Therefore, the contraction-relaxation cycle is obtained by to-and-fro rhythmic pumping of $Ca^{++}$ from its stores in the sarcoplasmic reticulum to myofibrils and vice versa.

All myofibrils are in a registered order and give the myocell its characteristic regular cross-striations histologically. However, cross-striations vary according to cell function. In relaxation sarcomere length is normally 2.4 μm while in contraction it is 1.5 μm. The systolic length of a sarcomere ranges from 1.86 to 1.95 μm and the diastolic length from 2.05 to 2.15 μm. The length giving maximal active tension, in relationship to the Starling phenomenon, has been calculated at 2.20 to 2.35 μm (end-diastolic reserve) (Spiro et al, 1968).

Different aspects of the sarcomere are revealed by electron microscopy during various phases of contraction and relaxation. In relaxation two clear "I" bands, formed only by thin filaments, are visible at both sides of a Z line. Internal to the I bands are two more dense S bands, which include both actin and myosin filaments. In cross section, one thick myosin filament is encircled by six thin actin filaments arrayed in hexagonal order. Other bands formed only by myosin filaments are visible in the central part of the sarcomere. They constitute the so called H-L-M complex which consists of a unique darker band in the center with two L bands and two H bands in lateral positions. Together the H-L-M complex and the S bands are defined as the A band. In normal maximal contraction, I and H bands disappear because of the total penetration of thin filaments on one side of the sarcomere into the other; in cross section, their number is double. The A band remains formed by the S and L-M bands plus the $C_m$ (maximal contraction) band, which includes both actin and myosin filaments (Fig. 8).

As indicated above the different types of bands and Z lines are clearly defined by electron microscopy. With light microscopy at high magnification, very thin Z lines may be recognized between two adjacent clear I bands when the myocell is relaxed. In contrast, Z lines become distinct in hypercontracted myofibers because of a drastic increase of their thickness.

## *Myocardial Damage Related to Myocell Function*

The myocardial cell may stop functioning in irreversible relaxation or in contraction or may progressively lose its force and velocity. We believe each of these situations produces a different morphologic form of irreversible myocardial damage.

## *Atonic Death in Irreversible Relaxation—Infarct Necrosis*

This type of myocardial necrosis is observed when myocells lose their capability to contract, becoming passive and extensible elements (Fig. 9). The loss of contraction both occurs and can be seen within a few seconds of experimentally occluding a dog's coronary artery (Tennant et al, 1935, 1936; Jennings, 1969). The acutely ischemic myocardium becomes cyanosed and because of intraventricular pressure shows a paradoxical, systolic, bulging. The histologic counterpart of this flaccid paralysis, with stretching and reduction in the thickness of the infarcted wall, is a thinning of the mildly eosinophilic necrotic myocells with elongation of sarcomeres and nuclei. These changes are visible in less than one hour of experimental coronary artery occlusion (Hort, 1968; Baroldi et al, 1977).

Other histologic changes in chronological sequence are seen in both experimental animal and man:

1. A centripetal polymorphonuclear, leukocyte (PMN) infiltration from the periphery of the infarct, occurs within 6-8 hours in the absence of or with a minimal edema fluid, fibrin and red cells. PMNs increase during the next 24 hours and disappear by lysis within the first week of their appearance, without evident destruction of necrotic myocells. Large infarcts may show a central area where the sequence of changes to be described does not occur. Rather, the mildly eosinophilic, stretched, dead myofibers persist. This is due to blockage of PMN penetration caused by maximal stretching of the central part of the dead tissue. Furthermore, if a marked PMN infiltration develops at the edge of the sequestered dead myocardium, the overall appearance may resemble an abscess with myocell destruction.
2. Fibrin-platelet thrombotic occlusion of intramural vessels included in the infarcted zone occurs parallel to, but not before the polymorphonuclear infiltration.
3. The healing process, which starts after one week, begins at the periphery by macrophagic digestion of necrotic material within sarcolemmal tubes and is followed by progressive collagenization. In contrast to others (Mallory et al, 1939), we and others (Barrie et al, 1957) believe the latter occurs without a granulation tissue response (Baroldi et al, 1975). Intimal obstructive thickening of small arteries is seen at the periphery of the early healing zone (Baroldi, 1967).

Three further findings complete histologic observations in this type of necrosis. First, the registered ordered of sarcomeres is maintained in remnants of dead myocells in healed infarcts

(>30 days) and, if entrapped, in scar (Fig. 9). Second, the lack of filling by postmortem injection of intramural arterial vessels is noticeable in an acute infarct (avascular area, see above, Fig. 10)). Third, this type of necrosis usually presents as one focus. It may affect the subendocardial zone or a greater width of the ventricular wall and can be transmural. In our experience its size ranges from less than 10% to more than 50% of total left ventricular mass. Very rarely it presents as small multiple foci in the subendocardium.

The commonly used designation "coagulation necrosis" given this form of necrosis, seems inappropriate because of the lack of coagulation of structures in its various phases. The more explicit term infarct necrosis seems more apt. Another comment relates to the hemorrhagic nature of infarct necrosis seen after fibrinolytic therapy (Fujiwara et al, 1986). Rarely, a myocardial infarct may be hemorrhagic e.g., when associated with wall rupture (Oliva et al, 1993) or therapeutic procedures. A third comment concerns "wavy fibers", i.e., undulated myocardial fibers as an early sign of myocardial ischemia by Bouchardy et al, in 1972. When found, their lack of specificity does not permit, per se, a diagnosis of ischemia. In fact, wavyness of normal myocells is usually observed around hypercontracted myocardial fibers (see below). Finally, it must be stressed that "ischemic" vacuolization and interstitial exudation or hemorrhage are not part of this type of injury.

## *Tetanic Death in Irreversible Contraction*

This form of myocardial necrosis presents an opposite morpho-functional pattern to infarct necrosis. Here the myocell is unable to relax and its function arrests in contraction, or more precisely in hypercontraction because of an extreme reduction in sarcomere length, much less than 1.5 μm calculated for normal contraction.

Several different morphologies result from such hypercontraction.

### Coagulative Myocytolysis or Zenker Necrosis or Contraction Band Necrosis

This injury has also been defined as anomalous contraction bands (Herdson et al, 1975); focal myocytolysis (Schlesinger et al, 1955); focal myocarditis with myofibrillar degeneration (Szakacs et al, 1958); infarct-like myocardial necrosis (Rona et al, 1959); myocytolysis with major contraction bands (Bloom et al, 1969); myofibrillar degeneration (Reichenbach et al, 1969, 1970) and contraction bands necrosis (CBN) (Ferrans et al, 1975). We prefer the term myocytolysis adding the adjective coagulative to emphasize the coagulation of contractile proteins seen. Alternatively, the term Zenker necrosis used in the past, for a similar change described in skeletal muscle (Adam, 1975), suffices. Two aspects can be distinguished: one which involves the entire myocell (pancellular lesion); the other limited to sarcomeres adjacent to the disc (paradiscal lesion). They are characteristic of and the only ones found following catecholamine infusion (Todd et al, 1985), also detected in many other pathologic conditions in man, e.g., pheochromocytoma, transplanted heart, thrombotic thrombocytopenic purpura, ischemic heart disease, malignant hyperthermia, scleroderma, etc. and in experimental models associated with catecholamine infusion, stellate ganglion stimulation, electric shock, magnesium or selenium deficiency, psychological stress, etc. (Baroldi, 1991) with its extreme example in "stone heart" (Baroldi et al, 1974).

#### *Pancellular Injury*

The first change is a hypercontraction of the whole myocell (Fig. 11) with markedly thickened Z lines and extremely short sarcomeres. Myocells become intensely eosinophilic and their sarcoplasm subsequently fragments into irregular total or partial transverse acidophilic bands or disrupts into diffuse granular material. These deeply staining cytoplasmic bands in hematoxylin-eosin sections alternate with clear, empty spaces or with spaces filled by small dark granules. Ultrastructurally, a transverse band appears as a small group of hypercontracted sarcomeres with highly thickened Z lines or as amorphous, darkly electrondense material, likely the result of coagulation of contractile proteins. The clear spaces are filled by normal or slightly swollen mitochondria that contain dense, fine granules and occasionally have ruptured cristae. The sarcotubular system is totally disrupted, while the basement membrane is essentially intact; only occasionally are interruptions seen in its continuity. Folding of the sarcolemma indicate the hypercontractile state of sarcomeres. Glycogen deposits disappear and there is no evidence of intracellular or interstitial edema. Blood vessels are not damaged and no associated hemorrhage with myocell necrosis and platelet aggregates or platelet-fibrin thrombi are found. (Todd et al, 1985). It seems likely that the degree of fragmentation of the rigid, inextensible myocells in irreversible hypercontraction is a consequence of the mechanical action of normal contracting myocardium around them.

The acute lesion described above is detectable in experimental conditions within 10 minutes of an intravenous infusion of norepinephrine or isoproterenol. It may involve a single myocell among thousands of normal ones or foci of a few myocells or large zones of myocardium. The degree of involvement in such experiments is dose-dependent and the lesion plurifocal.

This damage does not elicit a PMN leukocyte infiltration. Later, monocytes appear to digest necrotic material within sarcolemmal tubes leading to an alveolar pattern (Schlesinger's original "myocytolysis") followed by progressive collapse and collagenization by activation of interstitial cells (Schlesinger et al, 1955). This occurs in the affected areas without concurrent angiogenesis, i.e., evident granulation tissue formation. We believe this repair process is identical to that seen in infarct necrosis but have no exact idea of its speed in these generally smaller lesions in humans.

***Paradiscal Injury***

This myocellular lesion may be found in all conditions where coagulative myocytolysis occurs (Fig. 12). It presents a unique band of less than 15 hypercontracted sarcomeres adjacent to an intercalated disc. The remaining part of the myocell is normal. The band, with related scalloped sarcolemma, shows two typical ultrastructural aspects. One is a clear paradiscal contraction band formed by extremely shortened sarcomeres closely packed together with ill-defined, often fragmented, thin Z lines, while myofilaments are visible without evidence of rhexis. All mitochondria are squeezed together in the normal portion of the myocell. Another aspect is an increased electron density of different degrees of intensity, from almost clear to deeply dark, that crosses the whole paradiscal band. These dark bands are also visible histologically.

A paradiscal band is often observed at both sides of an intercalated disc and has, in general, a greater diameter than the other normal portion of the myocell. The adjacent normal myocells show a wavy disposition, possibly induced by the hypercontracted myocell. In cross section myocells affected by paradiscal lesions show large, deeply eosinophilic elements with a spoked-wheel aspect on PTHA stain.

We believe that paradiscal contraction bands are the equivalent of zonal lesions described in hemorrhagic shock (Martin et al, 1963, 1966). They are prevented by beta-blockers (Entman et al, 1965). The paradiscal band is observed within 5 minutes of intravenous catecholamine infusion. Its subsequent evolution is not known. Not seen at subsequent examination six days after onset, it may be a reversible lesion. Furthermore, we do not know if clear and dark bands are two separate entities or, more likely, sequential aspects of the same lesion. Since, in general, hypercontraction is characterized by thickened Z lines plus very short sarcomeres leading to a "coagulated" dark band, one may speculate that the dark aspect is the beginning of a hypercontracted paradiscal lesion while the clear one could be related to rebuilding of normal structure. The absence of fragmentation of myofibrils in this type of band is likely due to its paradiscal location at the extremity of an otherwise normally functioning myocell. On the other hand, segmental hypercontraction within a cell may lead to stretching and rhexis of adjacent non-hypercontracted sarcomeres, a finding often observed between two myocells in line: one hypercontracted and the other hyperdistended.

***Reflow Necrosis***

Frequently diagnosed as "infarct necrosis" clinically, reflow necrosis seems related to increased flow following ischemia in which catecholamines (Raab, 1970) and ionic calcium may have an important role. It may be defined as coagulative myocytolysis or contraction band necrosis associated with hemorrhage.

Reflow necrosis is found in patients subject to long-lasting resuscitative attempts or following heart surgery (Lie et al, 1978) and in experimental temporary coronary occlusion, and may be extensive. It may involve the inner half of the left ventricle and interventricular septum and be associated with massive hemorrhage (Fig. 13) producing concentric hemorrhagic necrosis (Gotlieb et al, 1977). There is also vessel wall damage with luminal platelet aggregates plus a scanty polymorphonuclear leukocyte exudate. In the experimental situation, reflow induces malignant arrhythmias (Reimer et al, 1977). Both the morpho-pathological changes and arrhythmias can be prevented in different ways, e.g., hypothermia, and by different chemical substances, including beta-blockers (Reimer et al, 1976).

***Contraction Bands at the Cut Edges of Living Myocardium***

We note that at a site of myocardial biopsy or along the cut edges of hearts excised at heart transplantation, living myocells retract with sarcoplasmic band formation (Fig.13). Experimentally, the depth of this hypercontracted margin is between 0.2 and 0.5 mm (Todd et al, 1985) and the pattern consists of hypercontracted sarcomeres with thickened Z lines, forming transverse and parallel bands without evidence of myofibrillar disruption. Such traumatic changes are not caused by ischemia.

## *Failing Death of Myocells or Progressive Loss of Function*

In this pattern and in contrast with the previous types of myonecrosis, the cell maintains its function with, however, a gradually reduced capacity to contract ending in an inefficient myocardial cell. The histologic marker is a progressive lost of myofibrils associated with intracellular edema and with different degrees of damage from mild vacuolization (moth-eaten pattern) to total disappearance of myofibrils. This produces an alveolar pattern but in contrast to other forms of myonecrosis mentioned above, the alveolar pattern lacks macrophages or any other associated cellular reaction. The impression is of colliquation or washout of myofibrils that leaves a sarcolemmal sheath with a "clear" alveolar appearance (colliquative myocytolysis) in its cytoplasm at most filled by edema and/or packed small granules (mitochondria) (Fig. 14).

## *Physiopathology of the Three Forms of Myonecrosis*

Each of these three functional forms of myocardial damage has a distinct structural and biochemical or better biomolecular nature. In irreversible relaxation intracellular acidosis displaces $Ca^{++}$ from troponin with loss of contraction (Katz, 1971/1972, 1988; Opie, 1993). In irreversible hypercontraction, intracellular alkalosis induces a rapid loss of ATP with a lack of energy to remove $Ca^{++}$ from troponin (Meerson, 1969) and/or a massive intracellular influx of $Ca^{++}$ (Fleckenstein et al, 1975) from increased membrane permeability. This leads, by activation of myofibrillar ATPase, to contraction and ATP consumption. One notes that after temporary hypocalcemia, restoration to normocalcemia induces myocardial contraction band lesions ($Ca^{++}$ paradox phenomenon: Zimmerman et al, 1967; Hearse et al, 1978). Perfusion with $Ca^{++}$-free blood following coronary occlusion protects the dependent myocardium (Ashraf et al, 1978).

In the failing death of a myocell there is a reduced capability of the sarcotubular system and mitochondria to bind $Ca^{++}$ (Bing et al, 1974), likely linked with local catecholamine depletion, with reduced intramyocardial cell $Ca^{++}$, loss of $K^+$ and increased intracellular $Na^+$. Myofibrillar lysis is induced by prolonged beta-blocking therapy (Sun et al, 1967), hypokalemia (Emberson et al, 1969) and hypocalcemia (Weiss et al, 1966).

### *Reversible vs Irreversible Myocardial Damage in Relation to Dysfunction*

Temporary or permanent and regional or global dysfunction of heart muscle (asynergy or dissynergy) is observed clinically. Three main patterns are distinguished viz. hypokinesis, i.e., reduction of contractility, akinesis or absence of contraction and dyskinesis, i.e., absence of contraction plus paradoxical systolic bulging. Radionucleide angiography, echocardiography and phase contrast magnetic resonance are essential to establish cardiac wall asynergy in life.

Echocardiography during dobutamine infusion can distinguish between permanent and temporary asynergic areas (Pierard et al, 1990). Of 314 akinetic segments in 33 chronic ischemic heart disease patients, 58% became normokinetic and 7% hypokinetic after venous bypass surgery. Dobutamine infusion was able to predict improvement in 198 of these 205 segments that recovered function after surgery (La Canna et al, 1994)

However, a need to establish the structural nature of cardiac dysfunction is paramount. Particularly because two different types of "viable", but noncontracting, myocardium have been proposed. They are: *stunned myocardium*, which occurs following reflow after a transient episode of ischemia produced by experimental temporary coronary occlusion and needs hours, days or weeks before contraction is restored (Braunwald et al, 1982; Ellis et al, 1983; Schwaiger et al, 1985; Stahl et al, 1987; Manyari et al, 1988; Gropel et al, 1990; Kloner et al, 1990, Moore et al, 1990, Taylor et al, 1992) and *hibernating myocardium* defined as "a state of persistently impaired myocardial and left ventricular function at rest due to reduced coronary blood flow that can be partially or completely restored to normal if the myocardial oxygen supply/demand relationship is favorably altered either by improving blood flow and/or reducing demand" (Rahimtoola, 1989). In other words in chronic ischemia the myocardium stops contracting teleologically, to save its structure ("smart heart"), and is ready to contract again as soon as ischemia is alleviated. Less clear, according to definition, is how an already hibernating myocardium can reduce its demands and yet return to function.

These two dysfunctional patterns (Bolli et al, 1988; Bonow et al, 1990; Narula et al, 2000) with an apparent diverging pathogenesis, i.e., reflow shortly after non-necrotic, acute ischemia in stunning and non-necrotic chronic ischemia recovered by reflow in hibernation, seem not to produce histologic signs or minimal and "reversible" ultrastructural changes affecting mitochondria in experimental stunned myocardium after 15 minutes of coronary occlusion (Kloner et al, 1998, 1989). By repetitive, brief coronary occlusions the stunning increases with an increasing number of occlusions ("sensitization"; Schroder et al,1988) and there is relaxation of muscle fibers (wide I bands), margination of nuclear chromatin, glycogen depletion, intra- and extracellular edema and marked alteration of collagen matrix components (cable, weaves, struts: Zhao et al, 1987).

Histologic and ultrastructural findings in transmural biopsies from dysfunctioning collateral-dependent areas in chronic angina patients with severe stenosis or old occlusion of one or more coronary arteries, with and without old infarct, demonstrated cellular swelling, loss of myofibrillar content and glycogen accumulation (Flameng et al 1987, Vanovershelde et al, 1993, Borgers et al, 1993). These changes, similar to those described above as colliquative myocytolysis were considered characteristic of hibernating myocardium and to be caused by repeated episodes of ischemia and not by chronic hypoperfusion (Vanovershelde et al, 1993); hibernating being considered an incomplete, progressive time-dependent degenerative adaptation to ischemia (Elsasser et al, 1997) with deterioration plus fibrosis (Schwartz et al, 1998).

However, flow response to dobutamine was markedly reduced in necrotic but not in hibernated and stunned segments, with an improved function in 11%, 16% and 55% respectively (Sambuceti et al, 1998). Interpreted as reversible "dedifferentiation" with "partial to complete loss of sarcomeres, sarcoplasmic reticulum, T tubules and abundant plaques of glycogen, strands of rough endoplasmic reticulum, lots of minimitochondria and a tortuous nucleus" rather than degenerative changes, the delayed functional recovery was imputed to slow resynthesis of the contractile apparatus (Borgers et al, 1993). On the other hand, in patients with chronic coronary heart disease and dilated cardiomyopathy undergoing heart transplantation, myocardial blood flow was similarly impaired in fibrotic and viable myocardium. This suggested that mechanism(s) other than myocardial fibrosis and coronary lesions determine blood flow impairment in end stage heart failure (Parodi et al, 1993).

In our definition of the three functional forms of myonecrosis, early changes were hyperdistension, enlarged I bands with normal Z lines in infarct necrosis, hypercontraction with very short sarcomeres and markedly thickened Z lines in coagulative myocytolysis or CBN and myofibril disappearance in colliquative myocytolysis. There is no method of establishing when these early changes are reversible or not. For example, in the first the disconnected interdigitations between thick and thin filaments might be reconnected if the main factors of stretching, i.e., intraventricular pressure and pulsation, are reduced; in the second, reestablishment of function might occur before the mechanical action of the normal contracting myocardium causes the disruption of hypercontracted myocells. In such case, thickened Z lines may represent an agglomeration of contractile proteins which might revert to normal. If so, the sliding theory of contraction should include the concept of a reversible "rolling up" of filaments at the Z line level; or if this is not the case, thickened Z lines should be a sign of irreversible damage. In the third lesion the cause of failure could stop with a rebuilding of myofibrils.

At present, there is no way to structurally recognize stunned or hibernated myocells or to relate those functional changes to the ones described in the previous paragraphs; especially considering that a similar amount of blood flow is present in normal and nonfunctioning myocardium in patients with unstable angina (Sambuceti et al, 1998; Gerber et al, 1999). On the other hand, when asynergic zones are matched histologically (Table 6), false-positive, i.e., asynergy with "normal" noncontracting myocardium and false-negative, i.e., myocardial "necrosis", even transmural, without asynergic segments were shown (Cabin et al, 1987). The apparent contradiction of a lack of asynergy associated with "transmural" necrosis can be explained by the definition given (transmural equals 75% of wall thickness). Perhaps, further quantitative studies are needed to establish the contractile status and type/age of the eventual, associated myocardial lesions. Is the stunned myocardium in (hyper)contraction and the hibernating one in (hyper)distension as suggested by the functional changes of these two types of damage? The clinical imaging of systolic and diastolic dysfunction should be integrated with the structural imaging of reduced or abolished capability of relaxation or contraction of myocells. Finally, one doubts whether in case of delayed recovery of a focal asynergic myocardium we are dealing with the time needed for the surrounding myocardium to hypertrophy and therefore compensate for the loss of function. There is no incontrovertible way to demonstrate the status of so called viable myocardium in vivo.

Besides CHD, asynergy may occur any time a critical mass of myocardium is involved by storage or parasitic diseases, etc. A peculiar structural change, "myofiber disarray" like that found in hypertrophic cardiomyopathy, deserves attention. The latter cardiomyopathy is an autosomal dominant disease with mutations in sarcomeric proteins—troponin T in particular (Yu et al, 1999)—considered "poison peptides" for the myocardial function, hypertrophy being related to a polymorphism of angiotensin I-converting enzyme without knowledge of the primary defect (Marian et al, 1998). Sudden death is a frequent event in this condition, possibly due to re-entry caused by disarrayed foci (Slade et al, 1993), which are absent in hypertensive cardiac hypertrophy (Takeda et al, 1999). Morphologically, disarray presents as an abnormal star-like disposition of hypertrophied myocardial cells joined by short, markedly hypertrophic myobridges associated with increased interstitial fibrosis (Fig. 14D); an architectural distortion which contrasts with the normal parallel alignment, needed for pump function. It provides for useless and endless increased contractility resulting in asynergic myocardium with its asynergic effect well demonstrated by the cases diagnosed clinically as restrictive cardiomyopathy who showed only a diffuse disarray associated with a heart of normal weight (McKenna et al, 1990; Baroldi et al, 1998). The etiopathogenesis and natural history of myofiber disarray we observed is unknown. Found in some specific regions of the normal heart, in hypertrophied hearts and at the periphery of myocardial scar and in other conditions, e.g., Noonans, Friedreich ataxia, lentiginosis, this architectural distortion shows a spectrum from a nonpathological pattern to a diffuse nonhypertrophic restrictive disorder or asymmetric septal (Teare, 1958) or apical hypertrophy or a diffuse hypertrophy seen in hypertrophic cardiomyopathy. The latter, being a combination of hypertrophic disarrayed myocardium and hypertrophic normal myocardium to compensate for the asynergy of the former; it is suggested that we should speak of "disarray cardiomyopathy" any time disarray becomes a pathogenetic factor (Baroldi et al, 1998). In the normal condition the few disarrayed foci may represent "nodal junctions" between muscle bundles at the site where they change direction. A sort of "centers of force" to help contraction. Similarly, around scar they may have a connecting function. In cardiac pathology, the presence, frequency and extent of this change are practically ignored—any attention being focused on hypertrophic cardiomyopathy. The latter frequently results in sudden death (Goodwin et al, 1976) so that myofiber disarray may be an important arrhythmogenic factor. Our investigation of this phenomenon was done by measuring its percentage (> 20%) with respect to the total histological area in all our material. A "pathological" disarray by our definition, was observed in 48% of sudden/unexpected coronary death, 46% of transplanted hearts, being absent in the cases that survived fewer than 7 days; 44% of intracranial hemorrhage; 26% of sudden/unexpected in silent Chagas' disease; 15% in cocaine abuse; 14% in congestive heart failure; 10% in AIDS, being absent in head trauma, electrocution and carbon monoxide intoxication groups. A maximal extent, i.e., disarray ≥ 20% in 3 to 8 regions, was seen in the first four mentioned groups. Disarray prevailed in the anterior/posterior left ventricle and anterior interventricular septum. No relationship was found with heart weight, coronary atherosclerosis, myocardial fibrosis, age and gender (data to be published).

## Cardiac Arrest

From a morphologic standpoint, structural changes related to cardiac arrest, particularly in CHD, are ill-defined. When the two opposite phases of the contraction/relaxation cycle are considered, two patterns can be recognized: one in myocellular relaxation and the other in myocellular contraction. On the other hand, from a clinical standpoint (Fish et al, 1985; Surawicz, 1985) the following electrocardiographic morphologies are associated with a cardiac arrest: (1) ventricular fibrillation as an end result of malignant arrhythmia; (2) asystole secondary to neurally-mediated bradycardia-hypotension (Milstein et al, 1989) or progressive reduction of the force-velocity of contraction; (3) electromechanical dissociation, i.e., loss of mechanical function (pulse, blood pressure, heart sounds) and consciousness despite a normal electrocardiogram (Fozzard, 1985), a pattern which, in general, ends in asystole associated with different conditions (e.g., pulmonary embolism, heart rupture with pericardial tamponade), even if no pathologic explanation can be found at autopsy (Hackel et al, 1993).

### *Myocardial Morphology in Ventricular Fibrillation*

In most CHD patients and in subjects dying a sudden/unexpected coronary death, ventricular fibrillation is the most frequent mechanism of cardiac arrest. Among many predisposing factors are an enlarged heart, mitral valve prolapse, myocarditis (Gradman et al, 1977; Pool et al, 1978; Proust et al, 1981; Pratt et al, 1983), psychologic factors (Engel, 1971), catecholamines or sympathetic overactivity, etc. but ischemia, particularly induced by plaque rupture and thrombosis is presumed the major one (Patterson et al, 1982; Willich et al, 1993).

Ventricular fibrillation has been defined as: "chaotic, random, asynchronous electrical activity of the ventricles due to repetitive re-entrant excitation and/or rapid focal discharge. Factors that enhance electrical synchrony facilitate, while factors that decrease electrical asynchrony hinder the development of fibrillation" (Zipes, 1975). The major difficulty is relating electrophysiological theories, e.g., reentry, abnormal automaticity, etc. to an anatomical substrate which is, in general, related to a myocardial infarct, or early ischemic changes often erroneously defined as contraction band necrosis. An "ischemic anisoinotropism" as an asynchronous contractile status, i.e., one contracted myocell amongst normally relaxed myocells has been proposed as a "dys-akinetic center" able to determine "micro-reentry" (Rossi et al, 1982,1985).

Morphological identification of the type of cardiac arrest could be relevant if we are to understand the mechanism of death, especially if sudden and unexpected. In reality, anatomical substrates for the pathognomonic electrocardiographic morphology of sudden cardiac arrest in general, and, ventricular fibrillation in particular, are still unknown. The latter is an obvious expression of an uncoordinated contractility leading to a rapid loss of pump function (Hottenrott et al, 1974). In its early phase myocardial blood flow increases intramurally because of progressive reduction of systolic compression of intramyocardial vessels ending in blood flow cessation. The heart nourishes itself so the chaotic, ineffective contraction starts a vicious circle with diminution of pump ejection inducing less myocardial nutrient flow which in turn reduces contractility. However, it is not clear what "uncoordination" means in terms of the involved myocardial elements. Does it simultaneously involve all myocells or bundles of cardiac muscles or different cardiac muscles? Is there any histo-morphologic equivalent of this sudden electrical storm? Any histological attempt to answer that question needs to discriminate pathological contractile status from changes due to rigor mortis. In the myocardium the latter starts one hour after death, resolves within 12-24 hours and may be absent in diseased hearts (Staemmler, 1961). However, little is known how this phenomenon behaves. Again, is this postmortem contraction simultaneous in all myocardial fibers or does it start in different cardiac muscles or bundles or cells?

In an experimental condition (normal hearts excised from anesthetized animals) using myocardial sampling at different time intervals, it was shown that myocell contraction (not hypercontraction with marked Z line thickening) starts at 40 minutes and rapidly extends and progresses thereafter (Vanderver et al, 1981) from the subendocardium (Lowe et al, 1983). Rigor mortis (sometimes named using the ambiguous and misleading term "ischemic contracture") is obviously paralleled by autolytic processes. Early separation of intercellular junctions and widening of intercellular space at disc level were occasionally seen.

In the ancient literature postmortem fragmentatio or fragmentation and segmentatio or segmentation of myocardial cells were mentioned. However, rupture of the myocell between the intercalated discs (fragmentation) or at the intercalated disc site (segmentation) is generally considered a non-vital artifact due to a microtome (Batsaki, 1968). Nevertheless, fiber segmentation shows an unexplained relationship to severe contraction of the myocells (Hamperl, 1929). On the other hand, prominence of the intercalated discs occurs prior to rupture of myocardial fibers in experimental extreme dilatation of heart chambers (Saphir et al, 1924, 1933). Finally, segmentation was interpreted as an agonal event, possibly related to ventricular fibrillation (Stamer 1907 quoted by Staemmler, 1961).

In most sudden and unexpected deaths associated with coronary atherosclerosis or in other "non coronary" conditions, e.g., subjects with a positive serologic test but without manifest Chagas heart disease or patients dying following brain hemorrhage or other cardiac diseases studied in our Institutions, all without resuscitation attempts, the histologic changes observed were (Fig. 15):

1. Bundles of hypercontracted myocells with thickened Z lines contiguous to bundles of hyperdistended myocells.
2. Widening of "stretched" intercalated discs between hypercontracted myocardial cells, often associated with segmentation. These hypercontracted myocells show square nuclei—due to contraction—rather than the usual ovoidal form.
3. Single or groups of hypercontracted myocells joined at their extremities with hyperdistended myocells. The latter may show partial or total sarcomere separation as a consequence of stretching and/or sarcomeric band formation.
4. Lack of eosinophilia and typical contraction bands seen in CBN.

Any one of these "ventricular fibrillation changes" may have a different extension and topographical distribution, from focal lesions in only one region to an involvement of the whole myocardium. They were never seen in more than 200 consecutive hearts excised at surgery for cardiac transplantation from patients with dilated cardiomyopathy, ischemic heart disease, valvular diseases or hypertrophic cardiomyopathy. This negative finding speaks against a technical artifact related to sampling and histologic procedures. Furthermore, to exclude the possibility that these changes may be secondary to ventricular fibrillation per se, we studied the hearts of ten anesthetized, open chest dogs in which ventricular fibrillation induced by electrical epicardial stimulus or intracoronary infusion of KCl was maintained for 30 minutes. In no hearts were there similar "ventricular fibrillation changes" (unpublished data). They were not described in an experiment in calves with circulatory support by ventricular bypass pump and in which ventricular fibrillation lasted from 1 to 40 hours (Ghidoni et al, 1969). On the other hand, they differ from those seen in autolysis or rigor mortis. The latter begins and resolves within 1-24 hours. Therefore, myobreakup should be present in all hearts at autopsy if rigor mortis was the cause.

In our opinion, the relationship between this myobreakup and ventricular fibrillation should be settled (Vassable, 1985). Segmentation and fragmentation could be artifacts produced at the site of structures already damaged in vivo. However, a metabolic disorder precedes ventricular fibrillation (Corday et al, 1977) and a beta-blocking agent was able to prevent the latter experimentally and in CHD patients. By intravenous catecholamine infusion in dogs monitored with an electrocardiogram, in no one instance did ventricular fibrillation occur, the only sign being S-T segment depression (Todd et al, 1985). Only when noradrenaline was injected into one coronary artery did both contraction band necrosis (without hemorrhage) and ventricular fibrillation develop (unpublished data). This means that local discharge of noradrenaline, e.g., medial neuritis of the atherosclerotic plaque, reflexes, etc., may only impair the syncytial rhythm resulting in instantaneous myobreakup. If this change expresses the morphology of an electrical instability, it may offer some indications how ventricular fibrillation works. People who have been successfully defibrillated may have only minor, local changes associated with rapid cardiac arrest or ventricular fibrillation following one premature beat or short run of tachycardia while unsuccesful resuscitation (protracted malignant arrhythmia) could coincide with diffuse myocardial damage associated with ventricular fibrillation. In all 15 patients in whom resuscitation was attempted with ECG monitoring, the cardiac arrest followed ventricular fibrillation associated with extensive myofiber breakup (to be published). A last question concerns how defibrillation by electrical shock works. By stimulation of epicardial cardiac nerves with diffuse intramyocardial release of noradrenaline?

## *Myocardial Morphology in Asystole*

An example of cardiac arrest with the myocardium in relaxation occurs when healthy people die accidentally from carbon monoxide (CO) intoxication because their myocells are relaxed. This was proved in ECG monitored rats killed by CO. The electrocardiographic pattern was of a progressively decreasing voltage, bradycardia, and asystole. In contrast, when reoxygenation was instituted, the animals showed foci of myocardial contraction band necrosis associated with arrhythmia and ventricular fibrillation prevented by a beta-blocking agent (Fineschi et al, 2000). Therefore, acute, severe hypoxia results in myocardial cell relaxation without any other change (vacuolization, edema, pathological contraction bands, etc). Only after reoxygenation does focal CBN occur, likely due to adrenergic overstimulation to prompt contractile function recovery. It is important to note that CBN was not a result of CO poisoning; a fact to keep in mind when resuscitating such victims. A further note is that reoxygenation results in coagulative myocytolysis without interstitial hemorrhage which characterizes reperfusion or reflow injury—showing a clear-cut difference between ischemia/reflow and anoxia/reoxygenation.

## *Other Myocardial Cell Injuries*

In other conditions injury to myocardial cells may result from, for example, parasites, e.g., trypanosoma cruzi, cryptococcus etc, often without evidence of any concomitant inflammatory or repair process, or from a storage disease such as hemosiderosis, glycogenosis etc, or from a myocarditis where leukocytes may kill myocardial cells, particularly in viral myocarditis where cytotoxic T lymphocytes seem to destroy the latter. However, without immunohistochemical recognition of the monocytic phenotype, it is impossible to discriminate between true cytotoxic lymphocytes and oligo-dendritic monocytes (Parravicini et al, 1991) as an expression of early repair of contraction band necrosis (Fig. 11 D), a possible reason for the diagnostic mismatch amongst cardiovascular pathologists (Shanes et al, 1987).

In defining the different aspects of myocardial cell injury, pathologists speak of cloudy swelling, hydropic or hyaline degeneration, vacuolization, etc. These are all nondiagnostic terms. The changes form part of one of the three types of myonecrosis described above. In general, a distinction is made between two forms of cell death. One, is oncosis (from Greek word "oykos") or ischemic death, i.e., swelling, vacuolization and blebbing ending in coagulation necrosis with karyolysis. The other, apoptosis, or physiological death is a cellular death that is genetically programmed and occurs in regenerating tissues. Its structural characteristics are a shrinkage of the whole cell and its nucleus ("half-moon, sickle nuclei") which fragments (karyorhexis) into "apoptotic bodies" within the dying cell or are extruded into the interstitium where they are phagocytosed by macrophages or neighboring cells ("cannibalism"). Oncosis involves all cells of one zone with a repair process ending in a scar, while apoptosis kills single elements here and there like lives dropping (ptosis=fall) from (apo) a tree, the cell disappearing without repair (Majno et al, 1995). The modish hypothesis is that several factors may trigger earlier the genetically programmed cell death: "suicide, execution or murder? "(Martin, 1993). An increasing number of reports (Colucci, 1996), based on indirect immunostaining demonstration by in situ nick-end labeling (TUNEL) technique able to detect DNA endings following nuclear rupture have been published. Indeed, apoptosis has been proposed to explain most, if not all, cardiovascular disorders. The process has been documented in smooth muscle cells and macrophages of the atherosclerotic plaque (Hand et al, 1995, Isner et al, 1995, Geng et al, 1995), in acute infarction (Bardales et al, 1996), reperfusion (Gottlieb et al, 1994) dilated cardiomyopathy and congestive heart failure (Katz, 1995; Narula et al, 1996; William 1999), arrhythmogenic right ventricular dysplasia (Mallat et al, 1996), myocardial stretching (Cheng et al, 1995), myocardial hibernation (Chen et al, 1997; Dispersyn et al, 1999; Lum et al, 1989), noradrenaline myotoxicity (Communal et al, 1998).

# CHAPTER 4

# Findings in Acute Coronary Syndromes

In keeping with the "unifying theory", presented by Gorlin et al 1986, CHD has a common etiopathogenetic denominator, i.e., rupture of an atherosclerotic plaque and consequent thrombosis and/or microembolization and/or spasm. These induce unstable angina, myocardial infarction and sudden death, all included amongst the "acute coronary syndromes". Thus plaque rupture becomes the center of the CHD universe, with rupture proposed to occur even in small, angiographically undetected plaques (Ambrose et al, 1988). Its prevention, and that alone, will resolve the CHD problem.

Our comparative pathological study was programmed to review very old concepts, i.e., whether occlusive coronary thrombosis causes an infarct (Hammer 1878, Herrick 1912, 1919) and whether thrombus plus embolization play a role in CHD (Chapman, 1974; MacIsaac et al, 1993).

The following review will compare findings in different groups of CHD and controls, outlining their meaning.

## Patient Data

### *Gender and Age*

Table 7 shows a significant prevalence of men in the sudden/unexpected death (SD) group (M/W = 7.0) and amongst healthy controls (M/W = 9.7). This difference was less evident among acute myocardial infarct (AMI) patients (M/W = 2.2). The highest incidence (35%) of sudden death occured in the sixth decade ($P < 0.01$) for men and in the eight decade (38%) for women ($P < 0.01$). AMI had a higher frequency amongst men (33%) in their seventh decade ($P < 0.05$) and amongst women (44%) in their eight decade ($P < 0.05$).

Both body weight and somatotype were within normal range in AMI, SD and control subjects.

### *Survival Time*

Survival time ranged from 6 to 12 hours to 30 days in AMI patients, from ten minutes (73%) to less than 1 hour (27%) in SD subjects. In AD cases death occurred instantaneously or within minutes or in less than six hours from carbon monoxide poisoning.

### *Activity and Cigarette and Alcohol Use*

About half of the SD subjects were engaged in an activity when they died suddenly (52 were at work, 44 walking and 13 driving). The other half were sleeping or resting. The distribution as far as the type of work performed was 60% manual (light 10%, moderate 22%, heavy 28%) and sedendary or executive work in 20% respectively. Of SD subjects, 28% were non-smokers, 22% mildly addicted (half a package per day) and 49% heavy cigarette smokers (more than one package per day). A similar distribution was seen in AD controls. Alcohol intake was

*The Etiopathogenesis of Coronary Heart Disease: A Heretical Theory Based on Morphology, 2nd ed,* by Giorgio Baroldi and Malcolm D. Silver. 

heavy (more than five litres per day) in only 13% of SD people and 10% in controls. Neither activity at the time of infarct onset nor cigarette and alcohol use were considered in AMI cases.

## Coronary Atherosclerotic Stenosis

All cases of this study had a physiologic intimal thickening that did not reduce the lumen of extramural, or subepicardial, coronary arteries. Pathologic intimal thickening seen in our material was atherosclerotic in nature.

### *Degree of Atherosclerotic Lumen Reduction*

Table 4 presents the degrees of luminal stenosis caused by atherosclerosis and the age distribution amongst 97 normal subjects. Thirty-eight of them (39%) proved to have a severe (≥ 70%) coronary artery stenosis; such stenoses being present in more than one main vessel in 16%. If the 74 subjects in this group who were more than 50 years old are considered, 46% had a severe luminal stenosis. These were at multiple sites in 19%. None of these individuals had clinical coronary heart disease or moderate/extensive myocardial fibrosis. One notes that 21% and 32% of these subjects had at least one <50% stenosis or 50-69% stenosis respectively.

Table 8 indicates the behavior of the variable "lumen reduction" in patients with acute infarction, sudden death or healthy controls in relation to their age. The frequency of severe stenosis from 40 to 69 years to ≥ 70 years age increased from 90% to 95% in AMI cases, from 39% to 52% in controls and remained stable at 81% in the SD group. The increase was not statistically significant. The AMI and SD groups were subdivided into patients with their "first episode" without extensive myocardial fibrosis and "2nd episode" or "chronic" patients with myocardial fibrosis greater than 10% of left ventricular mass. Furthermore, the findings in 100 noncardiac patients dying from various diseases of other organs, e.g., brain hemorrhage, pneumonia, liver cirrhosis, etc, were compared with AMI, SD,AD groups (Table 9). Data show a significantly greater frequency of coronary atherosclerotic obstructions demonstrable in patients with previous ischemic episodes in both AMI and SD groups. Furthermore, death may occur independent of the degree of lumen reduction and number of main arteries or branches with severe stenosis. It is significantly less frequent in patients with previous ischemic episodes. Of 200 consecutive infarct cases and 208 sudden death case, 72% and 64% respectively died at the first episode of illness.

Amongst all 455 patients with coronary heart disease, the left anterior descending branch in its proximal part was the vessel with the highest frequency of stenoses of any degree (90%) and critical stenoses with lumen/diameter reduction higher than 70% (41%) followed by the anterior segment of the right coronary artery (85% and 35% respectively ), proximal portion of the left circumflex branch (74% and 30%), distal portion of left anterior descending branch (68% and 29%), marginal (59% and 21%) and posterior (34% and 12%) segments of the right coronary artery. The vessels least frequently involved by any degree of stenosis were the left main trunk (50% all stenoses and 4% severe stenosis) and the posterior descending branch (10% and 3% respectively).

### *Length of Stenosis*

In all groups the length of mild stenoses (≤ 69%) was significantly shorter (≤ 3 mm) in AMI and SD cases while in chronic CHD the longest (> 30 mm) stenoses prevailed; amongst AD subjects, short and long stenoses had the same frequency. Severe stenoses (≥ 70%) generally had a significant tendency to increase in length with an increasing degree of lumen reduction. Amongst AMI patients, however, the majority of severe stenoses showed a significant shortest length (≤ 3 mm), in contrast to normal controls who mainly presented severe stenoses with the longest length ( ≥ 30 mm). It must be noted that along the course of the stenosis variations of lumen reduction exist.

### *Type of Stenosis*

The atherosclerotic plaque at the site of maximal lumen reduction was concentric in 70% and semilunar in 30% of all cases. In particular in AMI group it was concentric in 99% of the cases. Semilunar plaques showed a higher association with mild (60%) than severe stenoses (13%).

### *Morphologic Variables of the Atherosclerotic Plaque*

When morphologic variables in plaques with the same lumen reduction and intimal thickness were compared among the different patient groups (Table 10), significant divergencies were noted as follows:

1. Among AMI cases, atheroma, hemorrhage, calcification and lympho-plasmacellular infiltrates prevailed independent of the degree of lumen reduction. In contrast, these variables were significantly less frequent in healthy controls, while chronic ischemia and SD groups had an intermediate position. Chronic ischemic patients were more like acute infarct and sudden death cases than "controls".
2. Intimal hemorrhage was the least frequent variable found (14% of total mild and severe plaques). It was mainly observed in severe concentric stenoses located in a vessel related to an acute myocardial infarct.
3. Intimal or adventitial inflammation or both were present in all AMI cases, in the majority of chronic and sudden death cases and were significantly less in controls (Table 11). The association of this inflammatory process with proteoglycans (Table 12) was equal in all groups. Also in all ischemic groups the inflammatory reaction was present in most if not all stenoses independent of their degree and type in the same patient. In controls it was absent or found only in one or a few stenoses. In coronary heart disease groups, the inflammatory plaque reaction did not correlate with heart weight or extent of myocardial fibrosis, or with old thrombus. It correlated with acute thrombus, infarct necrosis, coagulative myocytolysis, and short severe stenoses. Furthermore, a significant high frequency of this inflammation was observed in atherosclerotic plaques of the ascending aorta in sudden and unexpected death people vs controls. In AMI and chronic ischemic patients the aorta was not studied.
4. A prominent and peculiar tropism of lympho-plasmacellular elements for adventitial nerves adjacent to the tunica media was noted (medial neuritis).

No morphological variable demonstrated any change in respect to the age, gender, or heart weight of patients, or in the extent of myocardial fibrosis, or coronary medial thickness. Table 13 presents a synthesis of the main significant variations in characteristics of atherosclerotic plaques amongst our different clinical groups.

### *Medial Thickness and Plaque Variables*

We found that maximal medial thickness ranged between 100 and 199 μm in the majority of vessel segments studied. It must be noted that medial changes were focally restricted to the region of atherosclerotic plaques only.

Medial thickness diminished significantly with both increasing intimal thickness and lumen reduction. This was particularly so with concentric plaques. The greatest medial thickness was associated with semilunar plaques and where lumen reduction was less than 70%. When maximal intimal and medial thickness were compared, irrespective of the degree of stenosis and type of plaque, both increased progressively until the intima was 2000 μm thick. With an intimal thickness greater of 2000 μm, there was an excess of both < 99 and > 200 μm width of media. In semilunar stenoses the media in the normal part of the vessel wall tended to be thicker than at the plaque site; in concentric plaques medial width was mainly uniform at any site, circumferentially. Only occasionally were both media and intima of lesser thickness. In 34 sections, most in acute infarct cases, the media was focally absent with an associated intense lympho-plasmacellular inflammatory reaction. No relation was established between medial thickness and morphologic variables in a plaque.

### *Heart Weight and Plaque Variables*

A pathological heart weight (≥ 500 g) was observed in 10% of control subjects; 43% of SD cases without extensive myocardial fibrosis; 76% of SD with extensive myocardial fibrosis; 39% of AMI without, and 53% associated with extensive myocardial fibrosis (Table 14). The number of both mild and severe stenoses was significantly higher in these heavy hearts versus "normal" ones. No relation was found between heart weight, gender, age and any other plaque variable.

## Coronary Occlusion

In healthy controls only one acute mural thrombus was found in a coronary artery. The affected patient had no clinical symptoms referable to it. Amongst coronary heart disease subjects the type of acute occlusive lesion found in subepicardial coronary arteries and branches was a thrombus. Its frequency was 15% in 208 sudden death cases and 41% in 200 acute infarct patients. Acute mural thrombi figures were 10% and 18% respectively. An acute occlusive thrombus was observed significantly less frequently in sudden death cases, but a significant excess of acute occlusive thrombi was seen in sudden death cases with extensive myocardial fibrosis (28%) and of acute mural thrombi in acute infarct cases without fibrosis (20%; Table 15). An old occlusive thrombus was present in 18% of AMI patients and in 6% of SD subjects. An old mural thrombus was seen in 4% AMI and in 1% SD cases.

In general, an acute occlusive thrombus was found in the infarct-related artery. It was located in the left anterior descending branch in 39, in the left circumflex branch in 11 and in the right coronary artery in 26 cases. In six cases more than one thrombus was found (LAD + RCA in five instances, LCX + RCA in one case). The left anterior descending branch was the main infarct-related artery in 52% of cases, the right coronary artery in 36% and the left circumflex branch in 11%.

We found the majority of these occlusive and mural thrombi in an area of severe (≥ 70%) luminal stenosis, that lesion being mainly concentric and longer than 3 mm. These acute thrombi were significantly associated with advential/ intimal inflammation, intimal hemorrhage, atheroma and calcification in the plaque. In contrast, old organized thrombi were related to a significant absence of morphologic plaque variables and associated with fibrous plaque.

Amongst AMI cases acute occlusive thrombi increased in frequency statistically with increasing infarct size, (see below), a behaviour not seen for mural thrombi (Table 16).

## Different Forms of Myocardial Injury in CHD

Our investigation revealed that the three forms of myonecrosis previously discussed are present in coronary heart disease and are often associated.

### *Acute Myocardial Infarct*

There was a myocardial infarct of different size in all 200 AMI cases included in our study.

#### Location

The infarct had an anterior or antero-septal location in the left ventricle in 39%, was posterior or postero-septal in 29% and antero-posterior in 32%. It involved the luminal third of the left ventricular wall in 23 patients (2 with occlusive thrombus) the inner two-thirds of the wall in 62 patients (26 with occlusive thrombus) and was transmural in 115 (54 with occlusive thrombus) (Table 17).

**Size**

Overall, infarcts ranged in size from less than 10% of the left ventricular mass to more than 50% with the maximum being 85% (Table 17). About half of all 200 fatal infarcts were small (less than 20%). Infarct size had a different distribution when acute infarctions occurred in apparently healthy people (1st episode) were compared with those that occurred in patients with chronic CHD. In particular, chronic patients showed a size less than 10% in half and less than 20% in 64% of cases. In AMI 1st episode the figures were 22 and 43 respectively (Table 18).

**Relationship to Coronary Artery Lesions**

The frequency of acute occlusive and mural thrombi related to infarct size is reported in Table 19. A significant correlation exists between the occurrence of acute occlusive thrombi and infarct size. In infarcts smaller than 10% the frequency of an acute occlusive thrombus was 20%, increasing progressively to a maximum of 86% with an infarct size of > 50%. Their occurrence in men (46%) was not significantly more frequent than in women (31%). Infarct size did not correlate with the number (Table 20) or degree and length (Table 21) of severe stenoses present in the whole coronary arterial system.

**Survival**

When the survival time of all 200 AMI patients was considered (Table 22), of those with small infarcts (<20%) 64% had a short survival period (< 2 days). In contrast, of those with large infarcts half survived more than 11 days. These significant findings ($P < 0.01$) were similar in both 1st and chronic patients, who also showed no significant divergency in age and gender distribution.

No relationship could be established between survival, infarct size and the frequency of acute occlusive thrombus (Table 23). In Table 24 infarct size is related to the main supplying artery or branch. The left anterior descending branch usually supplied the largest infarcts. It must be noted that in 37% of our cases, an infarct involved the adjacent vascular territories of vessels that were not occluded. No difference could be demonstrated between men and women with respect to survival time and the size of their infarct between survival time and heart weight, or between heart weight and infarct size. Despite a similar distribution of pathologic heart weight (≥ 500 g) in different decades of life, hypertrophy of the heart generally was significantly more frequent in men (69%) than in women (39%) in this AMI group.

**Heart Rupture**

Among 200 acute infarct cases 27 died of cardiac tamponade following rupture of the left ventricular free wall at the site of transmural infarct necrosis. In two other cases the rupture was located in the interventricular septum and in another five the left anterior (2) or posterior (3) papillary muscle had ruptured.

The majority (31) of these ruptures occurred in 1st episode AMI cases (Table 25). A significant higher frequency of rupture was observed in hearts with an infarct size of 21-30% of left ventricular mass (30%) followed by an infarct size of 11-20% (27%). In only three chronic cases was a rupture present. The percentage distribution was 21% for the 1st episode cases versus 5% for the chronic cases. This difference was significant. The frequency of an occlusive thrombus was 50% in the 34 hearts that had ruptured compared to 40% in 166 hearts that had no rupture. However, this difference was not statistically significant. No relation was found between the degree and number of coronary stenoses or heart weight and rupture. Amongst 208 sudden/unexpected death cases only one had ruptured the left anterior ventricular free wall. The small rupture, without cardiac tamponade, occurred in a zone with extensive coagulative myocytolysis. Of the 35 acute infarcts in this group none had cardiac rupture.

## Forms of Associated Myonecrosis in Acute Infarct and Sudden/Unexpected Death

### *Acute Infarct*

All AMI cases showed in continuity with the peripheral layer of infarct necrosis, but not in the subendocardial or perivascular myocardium, multifocal or extensive confluent areas of coagulative myocytolysis or contraction band necrosis of varying size. Furthermore, in most hearts (85%) the normal myocardium in the region of the infarct as well as in other areas of myocardium far removed from the infarct revealed isolated foci, that were sometime confluent, of this type of necrosis. The presence of this lesion could not be correlated with the degree of coronary damage, the size of an infarct or the presence or absence of an acute occlusive thrombus.

In 38% of infarcted hearts, colliquative myocytolysis was observed in the subendocardial and perivascular myocardium uninvolved by the infarct. The presence of this lesion too, did not correlate with the degree of coronary damage, the size of an infarct or the presence or absence of an acute occlusive thrombus. In Table 26 a synthesis of the different morphofunctional forms of acute myocardial injury in CHD is given.

### *Sudden/Unexpected Coronary Death*

The different types of myocardial necrosis and the extent of fibrosis found in relation to the degree of coronary artery lumen narrowing and the presence of acute occlusive thrombi in SD cases are presented in Table 27.

An acute infarct was documented histologically in only 17% of 208 sudden death cases, 12% in 133 1st episode and 25% in 75 chronic cases. The necrosis was extensive in eight, (> 20% of left ventricular mass) moderate (≤ 20%) in 16 and microfocal in 11 cases. The occurrence and extent of an infarct did not correlate with the degree and number of severe coronary artery stenoses. An acute occlusive thrombus was detected in a subtending vessel in 50% of 16 1st episode cases and in 16% of 19 chronic cases. All but two instances had the thrombus in an area of severe coronary stenosis. According to its histologic pattern the infarct was between 12 hrs and 30 days old. These SD cases, with associated acute infarction that was hours or days old but with no history of pain preceding death, clearly support the concept of "silent" infarct related to coronary heart disease (Cohn, 1989).

Coagulative myocytolysis was the most frequent form of myocardial necrosis found in SD cases. It was observed as the only acute lesion in 72% of cases and from 5% to 20% of different AD groups. In all but three of the latter the lesion was minimal while in SD hearts it was moderate-extensive in 29%. In most AD subjects and in about two-thirds of the SD cases, this type of necrosis was early; whereas in 13% of SD and 4% of AD cases it was alveolar while a healing stage was seen in 12% and 1% respectively. This means that the lesion preceded long before the sudden demise. Among 28 SD cases with normal or a coronary lumen reduction less than 50%, this necrosis was observed in 78%. In all 35 SD cases with infarct necrosis, CBN was seen at the perimeter of the infarct and in 83% elsewhere in the "normal" myocardium. No relationship was demonstrated between its presence and the extent and degree of coronary obstruction or the presence or absence of an occlusive coronary thrombus.

Minimal foci of subendocardial colliquative myocytolysis were observed in only 8% of the SD cases. All but two were in subjects with both pathological heart weights (≥ 500 g) and extensive old myocardial fibrosis. In AD cases no colliquative myocytolysis was found. No relationship was established between survival time and individual activity and the degree of coronary obstructive damage, type and extension of myocardial necrosis or heart weight (Table 27).

### *Myocardial Fibrosis Associated with Acute Myonecrosis*

In general AD subjects showed minimal fibrosis. In only 5% of cases was a single, small focus of fibrosis visible grossly. In contrast, fibrosis was minimal or moderate (< 10%) in 73% of AMI and 64% of SD cases and extensive (≥10%) in 27% and 36% respectively. Conversely, extensive fibrosis tended to significantly increase when an increasing number of coronary vessels showed severe stenoses. Recent myocardial fibrosis was seen in 31 SD cases, isolated in 5, associated with old fibrosis in 16 and old fibrosis + infarct necrosis in 10, the lesion being located in different areas. Most foci were minimal, only two being extensive and two median. Both recent and old microfocal myocardial fibrosis likely are a result of coagulative myocytolysis repair. In general, no correlation was found between the frequency of thrombus, incidence/extension of acute necroses and myocardial fibrosis and heart weight.

The different patterns of acute irreversible myocardial damage and fibrosis were seen more frequently in the left ventricular free wall, followed by the interventricular septum and then the right ventricle. Acute irreversible damage was not seen in the conduction system of SD cases and only in 8 1st episode and 10 chronic cases of the latter were microfoci of old myocardial fibrosis observed.

## Intramural Vascular Lesions

No heart in any of our studies, even in the presence of severe atherosclerosis of extramural arteries, showed an atherosclerotic plaque in an intramural arterial vessel.

### *Fibrous Thickening of the Intima*

The different types of intimal thickening described above in subepicardial arterial vessels were never seen in intramural arterial branches. In general, a fibrous intimal thickening affected small arteries surrounding or within scar tissue or adjacent to the annulus fibrosus or the membranous interventricular septum. It occurred without any difference in frequency between groups of subjects. In particular, intimal thickening of the arterial branch to the sinus node was present in 2% of SD and in 2% of AD cases; and in 10%, and 15% of cases respectively in the arterial branch of the A-V node.

### *Subintimal Hyaline Material*

This was observed as small nodular deposits in people older than 50 years. Only one or occasionally a few intramural arterial vessels (max 14 in a SD case) were involved. Exceptionally, this deposit appeared to stenose the lumen, the reduction being generally semilunar, and not exceeding 50%. Its exact nature was not determined (negative stain for amyloid).

### *Perivascular Fibrosis*

Perivascular fibrosis of intramural arterial vessels was another rare finding in the absence of myocardial fibrosis.

### *Atheromatous Emboli*

Only one atheromatous embolus was seen amongst the more than 14,000 myocardial sections of all groups studied. It was associated with a reactive intimal proliferation in an interventricular septal arterial branch. The adjacent myocardium was normal. The subject was a SD case who had one severe atherosclerotic stenosis of the anterior descending coronary branch and ulcerated atherosclerotic plaques of the aorta (Fig. 16).

### *Occlusive Arterial Platelet Aggregates*

Platelet aggregates could be demonstrated in the heart of SD cases as frequently as in controls (70% vs 76%) but they were infrequent in either sample. (Table 28; Fig. 16) No pathological changes of the vessel wall were noted associated with these aggregates. A significant relationship was seen between their frequency, particularly in the AD subjects ($P < 0.05$ for trend) and where a longer interval existed from onset of the terminal episode to death (Table 29).

No relationship was demonstrated between the presence of occlusive and mural thrombi in extramural coronary arteries or the presence of demonstrable infarct necrosis or contraction band necrosis and the frequency of arterial platelet aggregates. The relationship between the degree of lumen reduction, the frequency of cases and the number of intramural arterial vessels with platelet aggregates did not show any significant divergence between AD and SD groups. Finally, platelet aggregates were rarely observed within the conduction system being present in one instance in the sinus node of one SD case, and in the A-V node-His bundle of five SD and four AD cases.

### *Blood Stasis versus Platelet Aggregates*

Blood stasis was not demonstrated in 30% of total sections from the sudden death group and 40% of the AD group. Associated arterial and venous intramural stasis was seen in 45% and 40%, while venous intramural stasis alone was present in 24 and 19%, respectively. Arterial and venous intramural stasis was significantly more frequent ($P< 0.001$) in the SD patients with extensive myocardial fibrosis. In both SD and AD cases, where stasis was present, a single line of red blood cells or polymorphonuclear leukocytes or platelet aggregates were frequently seen layered in the vessels. The separation of blood elements was particularly evident in longitudinal sections of arterioles, the proximal tract being filled by red cells and the distal by granular material. Furthermore, adjacent cross-sectioned vessels—likely branches of the same stem—showed all possible combinations of these findings (Fig. 16).

The frequency of both arterial and venous platelet aggregates directly correlated with the presence and type of intramural blood stasis in all groups (Table 30). In the case of venous stasis alone, only venous platelet aggregates showed a maximal frequency.

### Medial Hyperlasia Obliterans

Medial hyperplasia obliterans (Fig. 17) was found in 52% of SD and in 78% of AD cases (Table 31). The higher frequency in the latter is statistically significant. The distribution of medial hyperplasia was greatest in papillary muscles, columnae carneae and the interventricular septum in all groups. A higher frequency of this vascular change was observed in the anterior papillary muscle than the posterior one of either ventricles. No relationship was found with the patient's gender and age. The frequency was practically the same in different decades of all subjects (Table 32). No relationship was observed between intramural medial hyperplasia, myocardial fibrosis, atherosclerotic obstructive damage of the subepicardial coronary arteries or heart weight.

## Chronic Coronary Syndrome and Congestive Heart Failure

Our distinction of AMI and SD cases into two groups, i.e., individuals who were normal when suffering their heart attack (1st episode subjects) and those who had an acute coronary syndrome during the course of a chronic CHD (2nd episode or chronic patients; see definition above) revealed divergencies in several parameters when they were compared between the two groups. The significance of these findings will be discussed in the Chapter 6. Here, the important focus is the difference between acute events which may happen at any time during the course of the disease and the other main outcome of coronary heart disease, namely congestive heart failure (CHF). The latter is generally interpreted as a consequence of repetitive acute, nonfatal, events which, associated with chronic ischemia, lead to progressive myocardial fibrosis and consequent failure of contractility. However, a solid morphologic background, in favor or against the many pathogenic theories of CHF (see below), is lacking. In order to establish the latter, we reversed the usual approach to study single diseases producing congestive heart failure. Rather we examined cases with the same clinical pattern of CHF independently of the underlying causative disease. Hence, with the same method of examination of the heart adopted in our comparative investigation, we studied hearts excised at transplantation for irrevesible CHF in consecutive patients 63 of whom had CHD, 63 dilated cardiomyopathy of unknown origin and 18 valvulopathy. The excised hearts had been arrested without emergency therapy or superimposed agonal events. Furthermore, control groups were compared with the CHF groups (Table 33). Since excised hearts are removed leaving the native atria in place, their weight was adjusted by adding the theoretical atrial weight (Reiner, 1968) according the following formula: actual heart weight 100/75 (Baroldi et al, 1998). The main results were:

1. The heart weight and transverse diameter were significantly ($p<0.0001$) greater in CHF hearts, while the left anterior wall thickness was similar to that found in control hearts (size/weight paradox), brain hemorrhage patients.
2. Severe single or multiple coronary stenoses (≥70% lumen-diameter) were present in all "ischemic" hearts and in few other hearts (Table 33).
3. Considering myocardial injury in the CHF group (Table 34), "silent" infarct necrosis was present in seven CHD cases associated with severe coronary stenoses. Two of these infarcts were transmural and five microfocal and subendocardial, all with an histological age of about 20 days. In the other groups a silent, transmural infarct 15 days old was present in one Chagas patient and a focal infarct of the anterior papillary muscle was found in 4 AIDS and in 1 brain hemorrhage cases. Contraction band necrosis, which showed different histologic patterns from early to healing stages, was observed in most of CHF cases but its extent $mm^2$ x 100 was minimal. Colliquative myocytolysis prevailed in the CHF group independent of the underlying causative condition, particularly associated with extensive myocardial fibrosis prominent in ischemic CHF patients. However, when the fibrous index (total fibrotic area/total histological area in $mm^2$ x 100) was calculated, even the ischemic CHF patients had an excess (> 80%) of histologically viable myocardium. Notably, collagenous tissue showed mainly an undulate aspect to its fibers in contrast to the dense, packed and linear makeup found in central part of an infarct scar. Frequently the scar tissue transformed in adipose tissue (Baroldi et al 1997; Fig. 18).
4. Focal interstitial lymphocytic infiltrates were relatively rare and small (less than 20 elements) in all groups with the exception of Chagas hearts.
5. The myocardial cells, both histologically and ultrastructurally, were not hyperdistended as, for instance, occurs in early infarct necrosis (Fig. 9).
6. Endocardial thickening due to endocardial thrombosis was a rare finding, while endocardial fibroelastosis starting with nodular smooth muscle cell hyperplasia followed by elastic tissue hyperplasia and ending in endocardial fibrosis, was present in most of these CHF patients and with a relative high frequency in controls, but normal head trauma people (Fig. 19)
7. No correlation was found between clinical and laboratory data and the extension and severity of all coronary and myocardial morphologic abnormalities ($r<0.30$ in all instances). Dimensional and functional clinical CHF findings were similar irrespective of the extent of fibrosis. No or mild heart dilatation was found clinically in 34% and 14% of ischemic and dilated cardiomyopathy CHF patients without any relation to morphologic findings, including colliquative myocytolysis (Baroldi et al, 1998)

# Chapter 5

# Revisiting Dogma Related to Coronary Artery Disease

Science evolves as a continuous turnover of hypotheses that require constant review/revision. It is time to reconsider each single caryatid which sustains the present conceptual temple dealing with the etiopathogenesis of CHD.

Readers will recognize that most, if not all, current etiopathogenic assumptions are only hypotheses and that they are mainly derived from clinical images of questionable interpretation. They will appreciate that, at present, clinicians deal with patients who have symptoms and usually have disease that is well advanced. Perforce, they do not often examine individuals when a pathologic process begins. On the other hand, while pathologists often study advanced disease at postmortem, they can investigate (a) evolving disease in persons dead at its different stages; (b) the meaning of common variables by studying other diseases and normal people dying from different type of accident; and (c) substantiate the structure of clinical images without which the latter remained unexplained. Also, at present, we have no experimental model reproducing the natural history of this disease.

The first caryatid to be revisited is the belief that functioning coronary collaterals do not exist (Helfant et al, 1970).

## Compensatory Function of Coronary Collaterals

In the natural history of CHD several, already mentioned, observations cannot be explained without the adequate compensatory function of collaterals. Since repetita iuvant (repetitions help understanding), particularly for this key point, we list again the main facts which prove this compensatory function.

1. Most patients at their first episode of CHD, 66% of noncardiac patients and 39% of normal subjects dying by accident had one or more severe coronary atherosclerotic stenosis. One may assume the vascular lesions had been present for months, if not years, without provoking any clinical history of CHD or producing a myocardial scar and with those individuals living a normal, if often stressful life.
2. No relationship between infarct size and number of severe coronary stenoses, as should exist if collaterals are absent, since more stenoses should mean greater ischemia resulting in a larger infarct.
3. The lack of correlation between the extent of vascular territory of the occluded artery and infarct size, with a note that the latter may extend into other territories nourished by a nonoccluded or nonstenosed artery; as is shown also in vivo by hypokinetic zones expanding in a region with adequate perfusion (Ahrens et al, 1993).
4. The presence of acute or organized thrombotic coronary occlusion without a related infarct.

All these facts and others (see above) support the concept that enlarged collaterals shown tridimensionally by casts, produce an adequate compensation for severe stenoses. A fact con-

*The Etiopathogenesis of Coronary Heart Disease: A Heretical Theory Based on Morphology, 2nd ed,* by Giorgio Baroldi and Malcolm D. Silver. ©2004 Eurekah.com.

firmed by the experimental occlusion of a severe stenosis without dysfunction, the myocardium being rapidly (within a few days) protected by a dramatic increase in collateral flow (Khouri et al, 1968). This body of knowledge seems sufficient to raise the question whether chronic ischemia exists.

We note that coronary cinenagiography (1) is unable to visualize intramural vessels and therefore collaterals; (2) has a limited power of resolution and 3) in interpretation conflict exists between radiopaque labeled blood flow competing with the nonradiopaque blood flow coming from other arteries. With these limitations it is not surprising that the technique cannot demonstrate collaterals. Yet, cineangiography is considered a gold standard clinically. One must ask when cineangiography will be able to demonstrate changes like those shown by plastic casts; a question posed by Mason Sones, the father of cineangiography, as long ago as 1968 (personal communication). On this subject, one notes how, by cineangiography, even in the presence of extremely severe or subocclusive stenosis, the radiopaque menstruum immediately fills the vessel distal to a stenosis without reduction—as we can judge from angiographic imaging—or delay of the radiopaque flow. Since the coronary injection is selective, the only route bypassing the stenosis is a plaque satellite anastomotic network (Fig. 20) and/or that provided by homo-intercoronary collaterals. The documented presence of these highly enlarged collaterals justifies speculation that: (a) they may participate in blood flow redistribution any time there are favorable pressure gradients; (b) induce changes in the amount and direction of flow in relation to any new obstruction in the connected system; (c) reverse flow any time an increased peripheral resistance ensues whether caused by coronary spasm or intramyocardial extravascular compression. Balloon inflation during angioplasty may suddenly occlude this satellite route, thus explaining, acute ischemia and pressure/flow reduction distal to a stenosis.

In coronary heart disease bilateral coronary ostial occlusion by atherosclerotic aorto-coronary plaque is an uncommon finding. This pathology was not observed in our material. Nevertheless, comment on the possible role of extracardiac coronary anastomoses, particularly from the bronchial arteries (Moberg, 1968) is opportune. Such anastomoses appear to be the unique source of blood supply in the presence of occlusion or severe stenosis of coronary ostia. Indirect evidence for the compensatory function of these connecting channels is given by cases with coronary ostial occlusions due to aortitis. We had opportunity to review 11 cases from the files of the Armed Forces Institute of Pathology, Washington D.C. Of these, five men and six women, with an age from 10 to 63 years and heart weight ranging from 200 to 720 g, only one died suddenly, two had microfocal subendocardial necrosis and four microfocal fibrosis. Not one had a history of ischemic heart disease.

A further comment relates to the collateral function in the most often used experimental model that of acute occlusion of a normal coronary artery in the dog. In our study of canine coronary arteries by plastic casts we demonstrated extramural inter-homocoronary anastomoses connecting epicardial branches on the heart surface at the time of injection. In contrast, in the pig, we did not find any extramural collaterals and, because of very strong and persisting postmortem contraction (rigor mortis), could not satisfactorily inject either intramural vessels or collaterals.

Extramural collaterals, not compressed by contracting myocardium, may give reason, in the dog, for a rapid redistribution of compensatory flow after the acute occlusion of a normal coronary artery. This anatomic and functional condition may also explain why after one hour of permanent occlusion, i.e., the time needed for all ischemic myocells to die, the resultant infarct affects only a small part of the territory supplied by the occluded artery. In that animal, the circumflex branch of the left coronary artery is always the dominant vessel giving origin to the posterior descending branch. When this vessel is ligated, the resultant infarct involves only the posterior papillary muscle and the postero-lateral subendocardial layer of the myocardium

(Jennings, 1969). In the dog collaterals seem to have an important compensatory role with rapid recovery from induced ischemia in most of the dependent myocardium.

A last point to be considered is the recanalization of an occlusive thrombus. This has been interpreted as an important source of distal flow redistribution (Snow et al, 1955), with thrombotic occlusion of these new channels proposed as a possible cause of death (Friedman, 1967). We believe that the occlusion of such channels formed in an occlusive organized thrombus in an area of stenosis already bypassed by collaterals, has no significance. For example, what is the direction of flow through them, to distal lumen or to adventitial vessels? Furthermore, the process of recanalization takes longer (Weisse et al, 1969) than it does for collaterals to develop. It is possible that such recanalizing channels provide a compensatory flow function when a new critical stenosis develops in the parent vessel with preexisting and functioning collaterals. This seems the case when they develop a well formed tunica media, an expression of increased blood flow.

## The Coronary Atherosclerotic Plaque

Rarely CHD may develop in the absence of coronary wall and/or luminal lesions but it occurs more often in association with atherosclerotic plaque with a differing degree and number of stenoses including nonfunctional (≤69%) ones. For instance, in acute myocardial infarct and sudden death cases a maximal stenosis less than 70% was found in 13% and 34% respectively (Table 9). Therefore, in their natural history, acute coronary syndromes may begin in the absence of a functional stenosis and, in the majority of cases, in the presence of old critical obstruction(s) compensated for by collaterals. What, then, is the meaning of the CHD/atherosclerosis association?

### *Active Coronary Atherosclerotic Plaque*

As any other pathologic process, the coronary atherosclerotic hyperplastic plaque shows vital changes which explain its natural history from beginning to end (Table 35). The concept of a plaque inducing "clinical activity" refers to specific changes within it that can trigger a clinical event. From the biomolecular viewpoint, activity coincides with some substances disrupting the fibrous cap leading to fissuration/thrombosis which causes an acute syndrome (Libby 1995, Newby et al, 1999). From the cineangiographic view point, lumen stenosis is the prime active determinant in chronic ischemia. Despite shadowy angiographic images, different morphologies are described, particularly in patients with unstable angina, e.g., "presence of luminal irregularity or haziness with ill-defined margins, a smudged appearance, inhomogeneous opacification within the lumen or changes suggesting ulceration or plaque rupture" (Cowley et al, 1989), with angiographic "evidence" of a coronary thrombus in 58% of patients with unstable angina, in contrast, to 5% of patients with stable angina. Fibrinolytic therapy improved both vascular imaging, marked by reopening of narrowed segments and attenuation of ischemic symptoms. However, symptoms and signs recurred in 71% of the latter patients (Gotoh et al, 1988).

Unfortunately, cineangiography fails to visualize all severe coronary lesions found at autopsy (Dietz et al, 1992). Angiographic findings cannot be precisely correlated with histologic ones. Postmortem coronary angiography imagings were compared with histology in 73 stenoses (ranging from 50% to 99% of luminal-diameter narrowing) in 39 patients dead after a myocardial infarct or following coronary by-pass surgery. The angiographic stenoses were divided into type I (smooth borders, with hourglass configuration and no intraluminal lucencies) and type II lesions (irregular borders or intraluminal lucencies). Of 35 angiographic type I stenoses only 11% presented histologic complicated plaques, i.e., those showing rupture, hemorrhage, superimposed partially occlusive thrombus or recanalized thrombus, the majority being histologically uncomplicated plaques, i.e., fatty or fibrous plaque with intact intimal surface and no superimposed thrombus. In contrast, of 38 angiographic type II stenoses, 79% were histologically complicated lesions (Levin et al, 1982). These postmortem findings indirectly support a previous cineangiographic report on both unstable angina and acute infarct patients in which type II stenosis was defined as "a coronary plaque in evolution precursor of impending infarction" (Ambrose et al, 1985). More recently it was found that angiographic "plaque rupture (irregular lesions) is a common mechanism for the progression of occlusive coronary disease but is not a mechanism whereby smooth walled plaques develop into more severe smooth walled lesions. Irregular lesions rarely become smooth lesions even after many years". (Haft et al, 1993).This proposal is difficult to accept if one considers "irregular lesions" as being synonymous with ruptured plaques. "The latter is not dependent on the occlusiveness of the underlying atherosclerotic plaque. ..only severely occlusive (≥90%) irregular lesions commonly proceed to occlusion (50% over a mean of 2,6 years) "(Haft et al, 1993).

The imaging of atherosclerotic plaque by other techniques such as angioscopy (Mizuno et al, 1992; Feyeter et al, 1995) or intravascular ultrasound imaging (Lee et al, 1994) or fast cine phase contrast magnetic resonance (Shibata et al, 1999) have not given any convincing contribution to a clinical diagnosis of an "active" plaque. In a review of invasive (angiography, angioscopy, intravascular ultrasound imaging per se or associated with elastography, Roman spectrography, etc) and noninvasive (magnetic resonance, nuclear scintigraphy, optical coherence tomography, contrast echocardiography) imaging techniques the conclusion was that the main determinants of plaque rupture, i.e., size of atheroma, thickness of fibrous cap and inflammation, are poorly evaluated (Pastercamp et al, 2000), as poorly as evaluated are other more important variables. However, one must make a point anytime a catheter is inserted in a coronary artery. Out of 408 CHD cases we studied (Table 8), the maximal stenosis found in a single case was less than 69% in 68, 70% in 67, 80% in 109 and 90-99% in 164 cases. This means that the majority of these stenoses had a mean residual lumen ranging from 900 to 50 μm (Fig. 7). How can a catheter, which has a diameter of approximately 1,500 μm cross such stenoses without breaking the plaque and forming a false lumen? This may result especially with ultrasound intracoronary technique. Shape and contour of a stenosis can be altered with a misleading higher frequency of semilunar stenosis with a competent lumen (Nakamura et al, 2001) providing an erroneous support to the concept of vessel wall remodeling following atherosclerotic plaque formation (Glagov et al, 1987). In 2121 coronary sections examined the plaque was concentric in 70%. The 30% of semilunar plaques only rarely showed a large lumen of "remodeling" shape. In particular in acute cardiac infarct cases the plaque in the related-artery was concentric in 99% of instances.

At present, only histology offers structural details of both vessel wall and intraluminal changes particularly when serial section studies of plaques are performed. For instance, in our experience serial sections of atherosclerotic plaques allowed us to see in the thickened intima small arterioles with a well developed tunica media. They were connected on one side with an intimal capillary-like plexus and on the other with an adventitial giant capillary network joined with arterial branches. Furthermore, the former directly communicated with the small residual lumen of the coronary artery. These vascular channels within and around the plaque may correspond to angiographic images erroneously interpreted as rupture or thrombosis (Fig. 20), particulary when an increased peripheral resistance with stasis in related plaque occurs (see below).

We have already listed facts which challenge the assumed relationship between stenosis and ischemia both in acute and chronic (Table 36) conditions. At autopsy, no differences with respect to the severity of coronary atherosclerosis were shown in various categories of patients with stable or unstable angina (Guthrie et al, 1975). The major difficulty is to recognize all dynamic factors linked with the acute coronary syndromes. At present we can only speculate about their sequence when looking at a complicated plaque, since we do not have an experimental model to reproduce the events nor can we follow it in humans. Any attempt to define plaque "activity" must consider all anatomical and dynamic factors recognized to this point. They include:

1. Luminal stenosis of any degree.
2. Satellite collaterals (homo- and intercoronary anastomoses), the anastomotic network around the plaque (connections between adventitial arterioles-capillary network and intimal vessels and residual main coronary lumen; Baroldi et al, 1967; Zamir et al, 1985) and the recanalized channels of an eventual organized thrombus.
3. Spasm of the coronary artery and the status of the tunica media.
4. Inflammatory reaction in the plaque, particularly its relation to the local nervous system
5. Vascularization/hemorrhage in the plaque.
6. The role of endothelial, smooth muscle, macrophage and mast cells in releasing growth factors or thrombogenic and/or vasoactive substances and replication of some of these cells in the reparative process.
7. Regional myocardial asynergy with increased intramural resistance (extravascular compression) and flow blockage in a related main subepicardial artery (increased wall stresses).

Emphasis has been given (Entman et al, 1993; Buja et al, 1994) to a marked "inflammatory" process mainly represented by macrophages, described in atherectomy material from patients with unstable angina or non-Q-wave myocardial infarction (Moreno et al, 1994) as well as the immediate site of a ruptured or eroded plaque with thrombosis in patients dying from acute myocardial infarction (Wal et al, 1994). In defining the inflammatory lymphocytic and plasma cell infiltrates in atherosclerotic plaques of ischemic heart disease patients we distinguished between a primary inflammatory process and secondary macrophage reaction to tissue injury.

## Coronary Occlusion

The second caryatid supporting the dogma on the common etiopathogenesis for the acute coronary syndromes relates to concerns the coronary occlusion. The assumption is that myocardial infarct, sudden death and unstable angina are caused by occlusive thrombosis following the rupture of an atherosclerotic plaque, a hypothesis supported by cineangiographic and pathological studies (Gorlin et al, 1985).

### *Angiographic Studies of Coronary Occlusion*

An angiographic total occlusion is defined as the absence of forward flow of contrast medium in an involved coronary artery while the angiographic equivalent of "a thrombosis is persistent staining of intraluminal material by the radiopaque menstruum, most frequently detectable in patients with total or subtotal (> 95% narrowing) occlusion" (De Wood et al, 1980); or "abrupt vessel cutoff with convex, irregular or ill-defined margins or (the) presence of contrast staining at (an) occlusion site, in association with release of occlusion or change in appearance at the occlusion site following intracoronary streptokinase or as (the) presence of intraluminal filling defects in relation to the occlusion site after patency was demonstrated" (Cowley et al, 1981, 1989). By coronary cineangiography, done in 322 patients with acute, Q wave (transmural) myocardial infarction, a coronary occlusion was seen in 87% of 126 subjects examined within four hours of the onset of their symptoms. In an additional 10% of patients a subtotal occlusion (≥ 95%) was observed. Similar results were obtained in that study amongst 82 patients evaluated between 4 to 6 hours of onset of their acute infarct (85% and 11% respectively). In contrast, amongst two groups each of 57 subjects, one examined between 6 and 12 hours and the other between 12 and 24 hours of an infarction, angiographic total or subtotal coronary artery occlusion was observed in 17% and 16% respectively (De Wood et, 1980). Another study of 341 patients with acute, non-Q-wave, myocardial infarction, in which

192 had coronary arteriographic studies within 24 hours, 94 between 24 and 72 hours and 55 between 72 hours and seven days after peak symptoms found total occlusion of the infarct-related vessel in 26%, 37% and 42% while a subtotal occlusion (≥ 90%) was seen in 34%, 25% and 18% respectively (De Wood et al, 1986).

A vessel cutoff was never seen in our study by plastic casts nor after injection of radiopaque material at postmortem. In all of our cases with occlusion with or without acute myocardial infarction, the vessel distal to an occlusion was always injected via collaterals.

## Pathological Studies of Coronary Occlusion

The major cause of coronary occlusion in coronary heart disease is a luminal thrombus and from the earliest reports of myocardial infarction, the attention of pathologists and clinicians focused on the frequency of occlusive coronary thrombus with the conclusion that at autopsy in the majority of large, transmural infarcts one is demonstrable (Chandler et al, 1974; Freifeld et al, 1983: transmural 91%, non transmural 51%). In contrast, in sudden coronary death studies the frequency ranged from 10% to 82% with a mean of 29%.

Here, a comment is appropriate on the often quoted pathological study supporting the relationship between plaque rupture and thrombosis (Davies et al 1984, 1986). In 100 unstable angina patients who died suddenly within 6 hours, an occlusive thrombus was found in 44% of them, at the site of a severe atherosclerotic stenosis. A "plaque fissuring" was present in 103 of 115 vessels showing either mural and occlusive thrombi, plaque fissuring being diagnosed as "a connection between intraintimal platelet-fibrin thrombus and the lumen that is demonstrable by the presence of injection media within the plaque" (Davies et al, 1984). First, these are cases of "expected" sudden death not comparable with sudden/unexpected death cases we studied with a frequency of an occlusive thrombus of 15%. Second, most of these patients had a myocardial infarct documented histologically and the 44% frequency of an occlusive thrombus is in agreement with the frequency found in our 200 acute myocardial infarcts. Third, one may question whether the source of injection media in the intima are the adventitial vessels connected with the intimal through intimal neovascularization rather than intimal rupture; injection media being an unreliable marker of plaque rupture.

Two other pertinent findings are: (a) fresh, large (≥ 3000 $\mu m^2$) thrombus characterized by its "layered organization, aggregates of platelets, fibrin and erythrocytes" was documented histologically in plaques removed by directional atherectomy in 44% of patients with unstable angina or recent (2 weeks) myocardial infarction versus 17% of patients with stable angina (Rosenschein et al, 1994) and (b) amongst 59 patients with definite angiographic features of an occlusive thrombus, all of whom subsequently had emergency surgical revascularization, a thrombus was recovered from 88%, by intravascular passage of a Fogarty catheter. The thrombi were "consistently situated proximal to the area of stenosis", and were described thus: "the leading edge of each recovered thrombus demonstrated varying quantities of acute inflammatory cells. The number of cells ranged from a few to several hundred per high-power field. The consistent feature of the distal part of every recovered thrombus was a thickened layer of fibrin and platelets. As the sections progressed toward the middle portion of the thrombus, fibrin and platelets became interspersed with red cells, creating a distinct layering effect" (De Wood et al, 1980).

In this context, the definition of the terms thrombus and coagulum is of paramount importance. For instance, in *Dorland's Illustrated Medical Dictionary* (25th Edition, 1974, WB Saunders, Philadelphia) a thrombus is defined as "An aggregation of blood factors primarily platelets and fibrin with entrapment of cellular elements causing vascular obstruction at the point of its formation" and a coagulum as: "a blood clot formed either in or out of the body". When there is slowing of blood flow, layering of all blood components occurs (Fig.16). As described so beautiful by Boyd in his pathology text (1965): "It is convenient to consider coagulation and thrombosis separately, although the two are usually inextricably combined. Coagulation or clotting can occur in the test tube or in the vessels after the blood has ceased to flow, as well as in blood which is still in motion. Its primary constituent is fibrin, in the network of which are entangled the various formed elements of the blood. Chief amongst these are the red cells, so that the clot or coagulum is red and soft, and is referred to as red clot or sometimes (unfortunately) as a red thrombus. A better term is a fibrin or coagulation clot. A true thrombus, as we shall see presently, consists primarily of platelets but these are associated with fibrin and a limited number of blood cells. It is correctly described as a white or firm thrombus. If the end result, the clot and the thrombus, may resemble one another, the process by which they are produced are entirely distinct". A thrombus is initiated by platelet adhesion at a site of damaged endothelium and factors from platelets (mainly), trigger recurrent platelet aggregation/release/fibrin deposition leading to layers of fibrin-platelet aggregates (Zahn's lines) formed without associated entrapped red cells and polymorphonuclear leukocytes. A thrombus presents in the gross as an opaque, gray-pink mass that is firmly adherent to the vessel wall and not easily removed from it. In contrast, a coagulum maintains the composition of the blood i.e., is composed mainly of red cells, a few leukocytes, platelets and thin strands of fibrin. Grossly it is a glistening, elastic mass, not attached to the vessel wall and easily squeezed from a vessel lumen.

Because of their entirely different structure these two processes may repair by different mechanisms. A thrombus undergoes organization by platelet contraction producing spaces in it that subsequently become endothelialized. At the same time capillaries sprout into it from the vessel wall and are accompanied by macrophages and fibroblasts with progressive collagen deposition. Healing ends usually by the occlusive thrombosis becoming a fibrous mass that fills the lumen and may show some recanalization.

Subsequent changes in a coagulum, comparable to these described above, are less well known. It seems likely that its completely different structure implies a different fate. It has been stated: "The average atheromatous abscess has been subjected to repeated micro-haemorrhage over a long period and its pultaceous contents is partly haemic in origin" (Morgan, 1956). In hearts excised at transplantation for cardiac failure due to previous myocardial infarct and where by-pass surgery had been done long before, we observed some venous grafts, with thickened but not atherosclerotic walls and with minor concentric obstruction filled by yellow, tooth-paste like material, easily squeezed from their lumen. Histologically, the graft along its whole course, had a lumen totally filled by atheromatous-like material. This finding suggests that a coagulum rather than undergoing organization like thrombus because it lacks a critical amount of fibrin, after a time may break down and transforms into atheromatous-like material giving the false impression of a "pultaceous occlusion" following hypothetical rupture of an enormous, nonexistent atherosclerotic plaque. There is no proof that a coronary thrombus may transform into "pultaceous" material as a coagulum may do.

A first criticism of the current dogma is that an occlusive coronary thrombus—a finding that only a pathologist can prove—is found in about half of infarct cases and in a minority of sudden/unexpected death cases.The belief that its absence at postmortem is due to its lysis is contradicted by the experimental observation that where a coronary occlusion is induced by intraluminal thrombosis followed by myocardial infarction, lysis of the thrombus did not occur. Rather, the vessel lumen remained totally occluded in 67% of dogs and partially reopened (75-25% stenosis) in 33% at 17 days (Weisse et al, 1969). We note that in this experiment thrombus formation was induced in a normal coronary artery. In contrast, in man occlusive thrombi are related to atherosclerotic plaques where a reduction of fibrinolytic activity of the vessel wall has been documented (Myasnikov et al, 1961). On the other hand, the spontaneous disappearance of a "coronary occlusion" in vivo is a common observation by cineangiography.

This disappearance is attributed to resolution of vessel wall spasm or to lysis of a thrombus but this may, or may not, be so.

The second criticism concerns the functional meaning of an occlusive coronary thrombus when present. Its frequency is related to infarct size and, in our experience, is minimal (20%) in infarcts less than 10% and maximal (86%) in ones greater than 50% of the left ventricular mass. This divergence may explain discordant reports in the literature in which a relationship between a correctly quantified infarct size versus occlusive thrombus is lacking. However, the frequency of a thrombus is only one variable. It may have very little significance when we have to establish its functional meaning and its cause-effect relationship with an infarct and/or sudden death. In our studies, beside its frequency being related to infarct size, the presence of an occlusive thrombus correlated with a severe (≥70%) degree of lumen reduction, the concentric shape of a plaque, its length, an atheromatous type of plaque and its lymphocytic and plasma cell inflammatory reaction. In other words, the variable "thrombus" is a multivariant phenomenon (Fig. 21). If one selects only very large infarct cases with luminal narrowing of more than 90% along the related artery and a concentric, long, atheromatous and inflamed plaque, the probability of finding an occlusive thrombus is 100%. This, however, does not prove that the thrombus caused the infarct. On the contrary, when one considers all dynamic aspects of a plaque, the hypothesis of secondary thrombus formation seems equally justified. A critically narrowed atherosclerotic plaque means that a functioning collateral system formed by satellite, homocoronary and/or intercoronary anastomoses and by a network of communicating channels around and within the plaque, by-passed the stenosed lumen (Fig. 21). This flow redistribution, with reduced anterograde flow counterbalanced by retrograde collateral flow distal to the stenosis, implies its hindrance within the tortuous residual lumen. Angiographically, to (during systole) -and-fro (during diastole) flow can be seen. This hemodynamic background may act, per se, with stasis of blood around and within the greatly vascularized plaque; or it may be associated with: (a) a mechanical action of the contracting myocardium on the coronary wall especially in exertion (Black et al, 1965); (b) coronary spasm and/or (c) extravascular compression of nonfunctioning myocardium with increased peripheral resistance and further blockage of flow in the related artery both in the residual lumen and in connected intimal/adventitial vessels. All are factors which may explain the sequence of the events in a dynamic or active plaque causing fissuring or rupture, hemorrhage mainly found after exertion (Burke et al, 1997, 1999), and thrombosis (Fig. 21). Bear in mind that plaque hemorrhage occurs mainly in the infarct-related artery at the site of reduced fibrinolytic activity by an atheromatous wall plus an increased coagulability any time there is tissue necrosis. In a patient who died within five hours of angiographically demonstrated coronary occlusion likely due to spasm, a mural thrombus at the site of a 70% stenosis without rupture was observed in a serial section study of a plaque. Clinically, the last episode was typical of an infarct but it was not demonstrated histologically because of a short survival time (Maseri et al, 1978). On the other hand, AMI patients show at 90 minutes a global flow reduction in both nonculprit and reopened culprit coronary arteries (Gibson et al, 1999).

The hypothesis of secondary thrombus formation is further supported by the frequent occlusion of a stenosis after surgical bypass grafting (Aldridge et al, 1971, Griffith et al, 1973). In a functional sense, the bypass flow is equivalent to a satellite collateral flow. On the other hand, if one accepts the concept of thrombus formation secondary to hindrance of distal flow associated with thrombogenic and/or vasoactive substances, the progression to occlusion of an already critical stenosis by subsequent mural thrombosis, is an event with little functional significance. In other words, and as has been shown experimentally by Khouri et al, (1968), the clinical angiographic aggravation of a critical stenosis or stenoses already bypassed by collaterals may not necessarily worsen coronary heart disease, in an ischemic sense.

On this subject, one notes an increasing number of angiographic reports (see above) theorizing a thrombotic occlusion at the site of a noncritical coronary atherosclerotic stenosis in an infarct-related artery. In studies comparing coronary angiographs before and up to a month after an AMI, authors found that "in the majority (66%) of subjects the myocardial infarction occurred because of the occlusion of a coronary artery that did not contain an obstructive (more than 50% diameter narrowing) stenosis on a previous coronary angiogram" (Little et al, 1988, Hackett et al, 1989)). However, in these studies the first coronary angiogram was performed a long time (mean 706 ± 685 days with a range from 4 to 2,298 days) before the infarct. Of 42 cases only four had coronary angiography three weeks before their AMI; three patients had "mild" coronary stenoses. Furthermore, there was no demonstration of an occlusion at the time of acute infarction. In a larger series of 283 low risk, medically treated ischemic heart disease patients, two angiograms were performed 4.6±0.1 years apart. At restudy 60 (21%) of these patients had developed a total of 75 new coronary artery occlusions and only 19% of them had a clinically recognized infarct. The majority (85%) of infarct-related coronary artery lesions were not hemodynamically significant (0-75% stenosis) at initial study (Webster et al, 1990). Nevertheless, the angiographer's viewpoint is: "In many patients (78%) who subsequently developed myocardial infarction, prior angiography revealed lesions that were < 50% occlusive in the infarct-related artery. Although the degree of narrowing in these arteries just before the onset of infarction was unknown, it was assumed that a more significant narrowing in the infarct-related artery had not slowly developed before the acute event. We suspect that this may have occurred because progression of coronary artery disease at restudy was uncommon in noninfarct-related lesions. Therefore, disruption of a mild or moderate atherosclerotic plaque with resultant thrombosis and total or subtotal occlusion probably explained the myocardial infarction. In patients with a previous normal appearing infarct artery, we assume that some degree of diffuse coronary disease was indeed present, but was not detectable by these angiographic techniques"(Ambrose et al, 1988).

This assumption that occlusive thrombi develop at noncritical stenoses seems weak for many reasons. First, it was never demonstrated at autopsy (Fishbein et al, 1996). Pathologists agree that where an occlusive coronary thrombus is demonstrated it forms, generally, at a site of critical luminal stenosis (≥ 70% lumen-diameter). This association was observed in our material in 93% of 200 fatal acute myocardial infarcts and in 100% of 208 cases of sudden and unexpected death with a thrombotic occlusion. In the six infarct cases with an occlusive thrombus in a stenosis less than 70%, the latter was in the range of 60-69% in two and 59-50% in six. All cases but one, had an infarct size greater than 50% (see above). In reviewing a series of 190 AMI patients, 117 associated with reopening of an occluded coronary artery by intracoronary fibrinolysis from papers of Cowley et al (1981) Ganz et al (1981), Mathey et al (1981), Reduto et al (1981), we found that 63 patients had an angiographic evaluation of the residual coronary artery stenosis following recanalization and the residual stenosis was critical in 84%. Second, the very high frequency of noncritical plaques even in healthy controls (Table 4) speaks against small plaques being prone to rupture and developing associated occlusive thrombi. If this was the case, this association should be seen frequently. Third, the previously mentioned angiographic studies were, in general, performed months or years before the infarct. Without knowledge of the "angiographic" degree of stenosis at the time of the latter, one can only defer to postmortem observations (Table 9). Furthermore, the concept of a persistent mild stenosis in an infarct-related artery (suggested by a lack of progression of stenoses in non infarct-related vessels) does not consider the effects on plaque progression caused by the regional-related asynergy. In the case-report from the Pisa Institute, (see below), after 12 months the left anterior descending branch and its vein graft (both normal at surgery) showed a severe lumen reduction due to atherosclerosis along their whole course. This raises the possibility that progression of atherosclerosis is related to dysfunction of dependent myocardium. Blockage or restriction, possibly recurrent, of intramural flow may enhance all conditions, i.e., physical, functional, neurogenic and biochemical from several cellular sources all of which may stimulate progression: the latter being much slower in vessels related to normally functioning myocardium. The higher length of coronary atherosclerotic plaques in chronic CHD patients supports this concept. As previously mentioned, another possibility is that the inflammatory reaction in the plaque may increase progression particularly in a radial (stenosing) direction. In our acute infarct cases the infarct-related active plaque was significantly shorter and severely narrowed.

A third criticism relates to the nature of acute coronary occlusion. In our review of arguments about causal factor(s) of coronary heart disease, we recalled a need to examine a phenomenon at its onset and during its development. All previously reported clinical-angiographic and postmortem studies deal with patients examined, at the best, within one hour of clinical onset of their acute ischemic syndrome or dying after a relatively long period, possibly with a variety of therapeutical maneuvers and drugs applied. Keep in mind that the passage of one hour is already a long period if primary events are to be distinguished from secondary ones.

At the Italian Institute of Clinical Physiology (Pisa) it was possible to follow a particular patient and all coronary angiographic and clinical events prior to and after a myocardial infarction he suffered during coronary angiography. Twelve months later, the patient had a heart transplant because of progressive, intractable cardiac failure and the excised heart was studied (Baroldi et al, 1990). This 45 year-old man had unstable angina pectoris for two weeks. A diagnostic coronary angiography was performed while he was clinically stable. The procedure showed both antero-septal and antero-lateral hypokinesis with a critical stenosis in the right coronary artery and two critical stenoses of the left anterior descending branch (LAD) one upstream and one downstream of the origin of the second diagonal branch. Aortic pressure was 137/70 mmHg and left ventricular pressure 130/12 mm Hg. The first ECG change (downsloped ST segment) without any subjective symptom or other clinical and angiographic sign was noted following the fourth LAD injection. Because of persistent ECG changes, another four LAD injections were performed without any changes in angiographic images or any other clinical subjective modifications. Only during the last injection did the postenotic tract of the vessel became fainter and disappear. Again, the image of LAD occlusion was not associated with other clinical and angiographic parameters or subjective symptoms. An intracoronary vasodilator and $Ca^{++}$ antagonist failed to restore blood flow. Following an intracoronary bolus (50,000 U) and intracoronary infusion of urokinase (10,000 U/min) for 20 minutes, ST changes and T waves tended to normalize with an image of recanalization. However, despite continuing urokinase infusion, the ECG changed again with LAD reocclusion. At this time, approximately 90 minutes after the first ECG ischemic changes, the patient felt mild chest discomfort. Percutaneous transluminal angioplasty was then performed successfuly with reopening of both proximal and distal LAD stenoses. Nevertheless, there was no benefit for the patient who experienced increasing chest pain and marked ST-segment elevation. Repeated contrast injection into the LAD demonstrated progressive disappearance of the vessel starting from its distal portion and extending to its origin from the left main trunk (Fig. 22). Since another balloon attempt failed to restore flow, the patient underwent emergency coronary artery by-pass surgery. The entire LAD was filled by an easily aspirated coagulum. Accurate probing documented a normal lumen without appreciable narrowing at any site. The LAD and implanted graft distended as soon as the clamp was released, but, as shown by a flowmeter, had no flow. Vessel and graft remained patent at repeated probing, but flow was never restored. The patient recovered from a large antero-lateral-septal infarct and was discharged home 15 days later. However, because of progressive, intractable heart failure without other episodes of ischemic heart disease he had a heart transplant 12 months later. The excised heart showed a massive antero-lateral-septal left ventricular scar (approximately 30-40% of the left ventricular mass) with an aneurysm of the antero-lateral wall. Multiple microfoci of fibrosis were detected in the remaining parts of left and right ventricles without evidence of any change in intramural vessels. The myocardium showed a diffuse loss of myofibrils (colliquative myocytolysis). The LAD and corresponding vein graft presented severe lumen reduction (LAD 90-95%, vein graft 70-80%) along their whole courses. An organized occlusive thrombus was found at the site of surgical anastomosis. The first part of the right coronary artery was occluded by an old organized thrombus in an area 90% stenosed by an atherosclerotic plaque. The left circumflex branch was mildly stenosed (50%) in its distal part. Atherosclerotic plaques in all coronary arteries revealed severe atheroma, calcification, and lymphocytic-plasma cell inflammation (medial neuritis) histologically.

As far as we know, this is the only case in the literature where it was possible to follow clinical events before and after a myocardial infarction and to have pathological documentation without superimposed agonal or resuscitative effects since the heart was surgically excised. The case demonstrates the following points:

1. The first ischemic ECG change occurred and persisted for 20 minutes, without angiographic evidence of coronary occlusion, chest pain or angiographic alteration of the LAD stenoses.
2. At 20 minutes angiographic occlusion was documented without chest pain or worsening of ECG changes or other clinical parameters. Only 70 minutes after coronary angiographic occlusion (90 minutes from the first ECG change) did both the ECG worsen and chest pain occur despite successful angioplasty.
3. A rapid sequence of up-to-date therapeutic interventions failed to restore permanent coronary blood flow. Only brief, temporary periods of reflow were documented during intracoronary urokinase (reocclusion despite continuous infusion of the drug) and after successful angioplasty. Paradoxically , ECG and chest pain worsened following the latter procedure.
4. The disappearance of the LAD lumen started from its distal portion and progressed retrogradely to its origin from the left main trunk.
5. Evidence at surgery of a patent LAD filled by an easily removed coagulum.
6. Pathologic documentation of a large infarct in the territory of LAD, corresponding to the hypokinetic zone diagnosed before infarction. Absence of an infarct in the territory of the right coronary artery which was occluded by a recanalized thrombus at the site of a severe stenosis. Absence of any type of changes such as old fibrin-platelet and/or atheromatous emboli in intramural vessels.

All of these facts indicate that in this case of unstable angina, the acute syndrome (infarction) started without documented angiographic occlusion with the latter demonstrated only 20 minutes after the first ECG changes. Therefore, the occlusion appeared to be secondary (no evidence of spasm) and, per se, did not promote any new objective or subjective signs for approximately 70 minutes. Furthermore, the nature of the angiographic occlusion was documented by inspection at surgery, and overall by its retrograde progression from the distal part of the vessel to its origin—and not at the site of stenosis—where the left circumflex branch had unrestricted flow. All data indicate that this angiographic occlusion was a pseudocclusion, namely an angiographic imaging of a meaningless secondary event due to flow blockage by intramural resistance with blood coagulation, not thrombosis, in the infarct-related LAD. There was no evidence of any type of intramural embolization, while "no reflow phenomenon" (Summers et al, 1971; Majno et al, 1967; Kloner et al, 1974; Gavin, 1983) seemed unlikely due to the short and rare periods of reperfusion and absence of malignant arrhythmias.

One may only speculate on the possibility of spasm in intramural arterial vessels (no increased blood flow noted after intracoronary vasodilator) or extravascular compression by dysfunctioning myocardium marked by worsening of the preexisting hypokinetic zone with increased stretching of myocardium by intraventricular pressure. However, questions arise. They include (a) how many angiographic occlusions in general and in particular how many of the 87% of acute infarct cases with total angiographic occlusion observed within four hours (De

Wood et al, 1980) or 70% within 6 hours (Rentrop et al, 2000) had the same type of "pseudocclusion" seen in this case; (b) how many "layered thrombi" of platelets, fibrin, and red cells recovered at surgery (De Wood et al, 1980) or seen in atherectomy material (Escaned et al, 1993, Rosenschein et al, 1994) or by intravascular ultrasound imaging (Lee et al, 1994) or that were suspected in angiographic imagings are only coagulated blood? Note that in this patient a large transmural infarct was fully established in a relative short time (20-70 minutes), despite instant and appropriate therapy; chest pain being an unreliable signal in timing the event.

Finally, one notes that an infarct may occur in the absence of a previous mild or severe stenosis (Eliot, 1974, Fineschi et al, 2001); and pseudocclusion (blood clotting because of increased intramural resistance) may explain "incomplete lysis of thrombus" in mild/moderate stenoses in acute infarct patients who have undergone intracoronary fibrinolytic therapy (Brown et al, 1986). On this subject, the frequent failure or inadequacy of acute recanalization following intracoronary fibrinolysis (about 50% of patients according to Brown et al, 1986 and 39% Vogt et al, 1993) may be due to the different results obtained according to the type of occlusion: pseudocclusion with easy lysis of a coagulum whether spontaneous or by fibrinolytic agents compared to true occlusion by a thrombus without lysis. Thrombi are present at autopsy in about 50% of the total population of fatal acute infarcts. We note that thrombolysis after acute myocardial infarction, in contrast to experimental temporary coronary occlusion, does not affect the prevalence of complex ventricular arrhythmias (Turitto et al, 1990).

### *Functional Coronary Occlusion or Equivalents*

The existence of coronary spasm was postulated long ago (Leary, 1935) and apparently documented by cineangiography of the main extramural coronary arteries in patients with angina pectoris (Dhurandhar et al, 1972; Oliva et al, 1973; Maseri et al, 1978) or acute myocardial infarction (Cheng et al, 1972; Oliva et al, 1977; Vincent et al, 1983). Contraction bands affecting smooth muscle cells of the tunica media of coronary arteries were proposed as a histologic hallmark of spasm (Factor, 1985). On the other hand, spasm or vasoconstriction of resistive intramural vessels is postulated to cause ischemia in humans (Hellstrom, 1982; Pupita et al, 1990; Maseri et al, 1992; Galassi et al, 1994; Sambuceti et al, 1997; Marzilli et al, 2000) and proposed in cardiomyopathic Syrian hamster as a cause of focal myocardial necrosis. At present, the histologic markers indicating spasm (Factor et al, 1982 and 1985) are equivocal and inconsistent. Meanwhile, the case with pseudocclusion presented above, raises the question whether the spastic occlusions observed at cineangiography are real; and if so whether primary or secondary. Furthermore, we do not know how long spasm may last and whether it involves the whole vessel or a segment of it; having in mind that (1) a cineangiographic occlusion, which shows a cutoff of the whole vessel, is defined as being caused by spasm when an intracoronary vasodilator reopens the lumen; and (2) the few autopsied cases with angiographic demonstration of "spasm" had the latter in a vessel with a severe stenosis (Maseri et al, 1978; Roberts et al, 1982), a fact confirmed by intravascular ultrasound of spasm at the site of focal plaque induced by ergonovine maleate (Yamagashi et al, 1994). Other functional mechanisms able to reduce coronary flow are the "steal syndrome" (Leachman et al, 1972) when collaterals steal flow from the territory of the parent vessel; or "infarct at distance" (Blumgart et al, 1940) when there is occlusion of the parent vessel; or perivascular fibrosis as a limiting factor of vasodilatation (Reagan et al, 1975).The first two mechanisms seem unlikely due to the extensive collateral network within the myocardium. The third mechanism is contradicted by rheumatic heart disease in which a diffuse perivascular fibrosis exists in the absence of ischemic signs. The role of spasm in coronary heart disease merits further investigation (see below).

Relative coronary insufficiency is often advocated in all conditions where a disproportion exists between myocardial metabolic demand and blood supply. A typical example is cardiac hypertrophy in which the fibrotic myocardial foci frequently found are interpreted as a consequence of nutrient flow insufficiency in respect of the increased size of the myocardial cell, inducing heart failure (Linzbach, 1947; Buchner, 1950). This argument will be discussed later (see chronic coronary heart disease-congestive heart failure).

## Small Vessel Diseases

Whenever there are inconsistent angiographic findings in the coronary arteries associated with an ischemic clinical pattern, the usual hypothesis is a blockage of small vessels by platelet aggregates or fibrin/platelet thrombi. This viewpoint was recently encapsulated in the following quote: "Without alternative explanation the differential diagnosis leaves embolization with microvascular obstruction as the leading suspect" (Topol, 2000). This unfounded hypothesis and the concept of microembolization from ruptured plaques (cholesterol emboli) or a proximal coronary thrombus (fibrin-platelet emboli) are other caryatids of the dogma to be discussed. When speaking of coronary "small vessel diseases" (James, 1967), we indicate diseases of intramural vessels which show totally different morphologic changes from those affecting extramural arteries including their smaller subepicardial branches.

The concept that platelet aggregates have a role in myocardial ischemia derived from the experimental intracoronary and intraventricular infusions of adenosine diphosphate (ADP; Jorgensen et al, 1967, 1970). The latter produced transient circulatory collapse, ECG ischemic changes, eventual ventricular fibrillation, a transient fall in circulating platelets due to aggregation in microcirculatory vessels and myocardial "infarct". Animals made thrombocytopenic or with platelets refractory to ADP did not show severe circulatory changes or myocardial necrosis. The conclusion was that ischemic heart disease may be caused by microcirculatory platelet aggregation secondary to ADP released from different sources, e.g., tissue injuries, lysis of red blood cells, etc. (Mustard et al, 1969) and particularly from erythrocytes damaged when crossing a stenosed atherosclerotic plaque (Brain et al, 1962); platelet adhesion being inhibited by adenosine (Born et al, 1964).

The only morphologic lesions found in human intramural arterial vessels that might possibly cause an acute coronary syndrome are aggregates of platelets and/or fibrin-platelet thrombi or emboli (Fig. 16). Already defined above (Tables 28, 29) such findings are reported often in ambiguous terms and without a clear-cut distinction whether they formed in situ or embolized from proximal sources. Few postmortem studies on sudden death (Table 37) demonstrate these lesions; several angiographic reports on AMI ischemic heart disease patients with normal coronary arteriography take for granted that this pathogenic mechanism exists.

Reviews emphasized the complexity of (1) the regulation of vascular resistance in vessels with a diameter greater than 100 μm and include several factors such as, autoregulation, $O_2$ consumption, sympathetic stimulation, serotonine, adenosine, vasopressine (Marcus et al, 1990); and (2) of platelet-endothelial interactions and the seeming ability of some diseases such as atherosclerosis, diabetes mellitus, hypertension, uremia, hypercholesterolemia, preeclampsia, to impair release of endothelium-derived relaxing factor and nitric oxide (vasodilatation) and anti-aggregatory platelet factors (anticoagulation and fibrinolysis) from one site and to promote the endothelial release of vasocontrictive substances and platelet aggregating factors from other sites (Ware, 1993). Experimental coronary constriction of 60-80% in the dog determines a cyclic blood flow reduction secondary to transient obstruction of the stenosed vessel. It is not clear if this transient obstruction is due to platelet aggregation or spasm (Folts et al, 1982) to both, or something else.

Platelet adhesiveness can be stimulated by several factors, not necessarily linked with an atherosclerotic plaque; for example, catecholamines (Bridges et al, 1966) or endothelial derived factors are important. The question is whether an obstruction caused by platelet aggregates or the release of vasoactive amines from them with resultant local vascular spasm could explain ischemic episodes and trigger sudden death. When reported (Table 37), the number of vessels occluded by platelet aggregates in cases of AMI is small. In our observations there was no difference in this finding between SD cases and normal subjects. On the other hand, in human pathology some conditions exist which can be defined as "experiments" of nature. Thrombotic thrombocytopenic purpura is one. This is an unique disease marked by a diffuse occlusive microangiopathy of intramural arterioles (Fig. 23) plus diffuse platelet aggregation in normal arterioles. Furthermore, it is characterized by other anoxic factors such as severe hemolytic anemia—and therefore an increase of ADP—hemorrhages or neurologic disorders, including convulsions with increased cardiac activity. Not one of the 39 cases we studied and 220 cases reviewed from literature, had symptoms or signs of ischemic heart disease or died suddenly. In 31% microfoci of CBN were observed. Similarly, in 53 cases of sickle cell anemia, the plugging of small vessels by sickled erythrocytes (documented in vivo; Knisely 1961) was never associated with myocardial damage of any type (Fig. 23; Baroldi et al, 1967; Baroldi, 1969). One also notes a normal left ventricular performance and an absence of ECG changes or MB iso-enzymes of creatine phosphokinase during sickle cell crisis (Val-Mejias et al, 1974). A negative finding which also speaks against the "disseminated intravascular coagulation" proposed as an ischemic factor (Hardway et al, 1961; McKay et al, 1965), at least within the myocardium.

The other aspect is embolization of platelet or fibrin-platelet masses from an occlusive or mural thrombus in the main artery supplying an ischemic area. This finding seems peculiar to patients with unstable angina who die suddenly (and "expectedly") within 6 to 24 hours. In 80% of these patients an infarct has been documented histologically (Falk, 1985; Davies et al, 1986 (Table 37). It is difficult to determine how many such vascular lesions are microemboli rather than fibrin-platelet thrombi formed locally in intramural vessels within the infarcted myocardium as a secondary phenomenon. In fact, they are never seen in the normal myocardium around an infarct. A progressive blockage of flow (avascular area) by the stretched, necrotic myocardium plus local factors such as neutrophils, wall degeneration, etc, could explain secondary thrombosis in situ. Furthermore, such emboli have been described in AMI patients treated with coronary thrombolysis (Waller, 1987, 1991); and one may ask if other therapeutic procedures, e.g., resuscitation attempts, induce embolization. Terminal therapy was not reported in most pertinent studies. One can not exclude spontaneous platelet emboli from a related ruptured plaque of a coronary artery. The question is whether these emboli are the cause of "microinfarcts" leading to sudden death. The number of occluded vessels is not reported or when reported is astonishing low. For instance, of 260 sections from zones perfused by a thrombosed artery only 72 microemboli were found in 29 sections (Falk, 1985), a finding which raises doubts about a claimed cause-effect relationship. More important, is that the associated myocardial necrosis called a "microinfarct" has clear-cut histological features of microfocal, often confluent, contraction band necrosis. We stress that this non-ischemic lesion is caused by adrenergic overstimulation and platelet aggregation has never been demonstrated in the early phase of this lesion, following experimental infusion of catecholamines either by electron microscopy (Todd et al, 1985) or by Cr-labeled platelets (Moschos et al, 1978).

The claim that an increased influx of neutrophils follows reinstitution of flow into an ischemic area by thrombolysis, surgical bypass or angioplasty may obstruct small vessels by plugging and/or by vasoactive substance released from polymorphonuclear leukocytes (Engler et al, 1986; Dreyer et al, 1991; Entman et al, 1991 Mazzone et al, 1993, Entman et al, 1993) has no basis in pathologic findings. In the natural history of coronary heart disease neutrophils appear 6-8 hours after the onset of an infarct, when this lesion is fully established. There is no evidence that these leukocytes aggravate the lesion. Neutrophils are not part of atherosclerotic plaque inflammation, are only seen in infarct necrosis and are rare in the reflow necrosis; a finding not observed in human acute coronary syndromes. Experimentally, infarct size was reduced by selective inhibition of neutrophil cytotoxic activity affecting neutrophil migration into injured myocardium (Amsterdam et al, 1993); however, neutrophil depletion did not protect against reperfusion damage (Carlson et al, 1989).

The third aspect is embolization of atheromatous material from a ruptured plaque. In contrast to spleen, kidney, brain, etc, where cholesterol emboli are often seen, in more than 14,000 myocardial sections of all our groups, only one atheromatous embolus was found in a small intramural arteriole (Fig. 16).

Practically, in the natural history of coronary heart disease other types of small vessel diseases do not exist. Platelet aggregation or embolization, even in a subset of patients with unstable angina, seems an unlikely ischemic factor, particularly if one considers that 38% of unstable angina patients are hypersensitive to spasmogenic stimuli with respect to acute infarction (20%) and stable angina (4%) (Bertrand et al, 1982). Spasm could induce plaque rupture plus embolizing thrombus. Moreover, no beneficial clinical effects occurred in several studies using thrombolytic treatment in unstable angina (Neri Serneri et al, 1992).

A final comment relates to distal embolization of atherosclerotic plaque material and/or thrombus formation during balloon angioplasty (Saber et al, 1993). The major mechanism of dilatation by this procedure is plaque fracture (Waller, 1991). This should imply microembolization of pultaceous or eventual thrombotic material as well as thrombus formation. In these cases, distal embolization of atherosclerotic (29%) thrombotic (49%) or mixed (17%) material was observed in a few intramural vessels (mean number 3.9; Waller, 1991). In experimental angioplasty in normal swine producing eccentric medial disruption and formation of a crater, a thrombus was totally absent (Gravanis et al, 1993). The latter was also absent in human plaques fractured by angioplasty (Wanibuchi et al, 1992). Furthermore, successfully dilated coronary lesions with an angiographically visible dissection are no more likely to develop restenosis and are not associated with a worse clinical outcome at six-month follow-up than are dilated lesions without visible dissection (Hermans et al, 1992).

Sudden death or myocardial infarction is a rare event following balloon angioplasty (4-5% acute complications). This may indicate that embolization following plaque rupture may have little, if any, functional significance. On the other hand, the rarity of atheroemboli in sudden death/coronary heart disease cases without emergency or invasive procedure, including postmortem injection of radiopaque material, confirms that spontaneous plaque fracture is a rare event and secondary to other mechanisms such as spasm and/or stasis due to increased peripheral resistance. These mechanisms may prevent embolization.

In summary, when small vessel disease exists, our findings suggest it is not associated with ischemic heart disease and in all our studies we found no case which could be explained by intramural pathological changes, including the rare cases of acute myocardial infarction with normal coronary arteries (Eliot, et al 1974; Fineschi et al, 2001). In the literature, obliterative intimal thickening observed in arterioles of the conduction system ("fibromuscular dysplasia", see below) were linked to sudden death. This change is also observed in healthy controls and there is no proof that it induces acute coronary syndromes, i.e., sudden death. Also found in papillary muscles and columnae carneae, this morphology can be observed in endomyocardial biopsies where, erroneously, it may interpreted as small vessel disease and a cause of ischemia. Nevertheless, in our experience patients with unstable angina who died suddenly ("expected sudden death") may occasionally show contraction band necrosis of different age, intramural fibrin/platelet thrombi and mural thrombi (source of the previous one?) in an extramural main artery. For instance, a 33-year-old man with episodes of unstable angina, died suddenly of cardiac arrest in ventricular fibrillation during angioplasty. At postmortem, there was an occlusive, recent (> 3 days) thrombus superimposed on a 80% stenosis of the RCA, with 80% stenosis of LAD, 50% of LM and LCX, all without evidence of both occlusive and mural thrombi in the latter vessels. There was no infarct, while 219 foci and 272 myocells x 100 $mm^2$

with CBN of various age were present in anterior left ventricle, 594 and 461 respectively in posterior left ventricle, 19 and 36 in the interventricular septum and 6 and 22 in the right ventricle. No topographical relation was found with the number of arterioles occluded by fibrin/platelet thrombi (9 in the left ventricle, 16 in the interventricular septum). All atherosclerotic plaques had medial neuritis. Further observations will help us in understanding these (rare?) cases. At present, the dilemma is a) whether these platelet/firbin thrombi are emboli from extramural coronary thrombi or form in situ as result of catecholamine overactivity; b) their role in SD. Is it a secondary phenomenon or a prime one that causes CBN and ventricular fibrillation?

## Different Forms of Myocardial Injuries

The concept that myocardial necrosis equals myocardial infarction with a latter a singular entity in ischemic heart disease is another pilaster which supports current dogma. However, the complexity of coronary heart disease increases when the different patterns of myocardial necrosis found in CHD patients are recognized.

Of the three forms of morpho-functional myocardial damage described above (Table 26), each has its own morphology. There seems little possibility that each has the same pathogenesis, consequently each must be caused by a specific biochemical disorder (see above). Our reasoning is that any distinct pathologic entity must have its own cause and pathogenetic mechanism; the classification of diseases is based on this rule. Many causes may present a similar nosologic pattern but a pathogenetic mechanism always results in a constant morpho-functional disorder. Thus, infarct necrosis is apparently the result of a sudden nutrient flow reduction, coagulative myocytolysis or CBN is likely due to adrenergic stimulation or any other factor acting through free radical-mediated lipid peroxidation (Mak et al, 1988, Hori et al, 1991; Fineschi et al, 2001); and colliquative myocytolysis is likely linked with catecholamine depletion with reduced intracellular $Ca^{++}$, loss of $K^{+}$ and increased intracellular $Na^{+}$.

The conclusion, therefore, is that the "metabolic" and "ion" theories of myocardial cell death (due to massive cytosolic calcium overload with inhibition of glycolysis and lack of provision of glycolytic ATP"; Opie, 1993) in CHD should consider these different forms of myocell death. To speak of "ischemic" contraction band necrosis or use, in a vague sense the term "myocytolysis", as an expression of nutrient flow reduction, seems incorrect. Furthermore, the association of these different patterns of myocardial impairment forces us to add other pathogenic mechanisms to explain the nature of acute coronary syndromes and, in particular, its cause and complications.

### *Acute Myocardial Infarction*

Often, experiment may divert the mind from the reality of human pathology. From the earliest models of acute coronary occlusion in dogs and rabbits producing myocardial infarction and/or sudden death (Cohnheim et al, 1881), countless similar experiments have been done. In particular, data obtained mainly from dogs have influenced our thinking on human myocardial infarction. Already we noted that the canine coronary vessels differ from those in humans and other animals because collaterals join extramural arteries and their branches. These superficial collaterals are able to function since they are not compressed, as are the intramural ones, by the bulging of ischemic myocardium which occurs within a few seconds from occlusion. This explains why only a small subendocardial-posterior papillary muscle infarct of the left ventricle is established in the dog within one hour of permanent occlusion of the left circumflex branch (Jennings et al, 1969). Reopening of the occluded coronary artery after 20 minutes prevents infarction, while temporary occlusion of different duration (40 minutes, 3 and 4 hours) followed by 2-4 hours of reperfusion or permanent occlusion of 24-96 hours results in the "wavefront phenomenon", i.e., a progressive transmural expansion of the myonecrosis from endocardium to epicardium. Within this evolving process three different zones were described: (1) a "central" one with myofibrillar relaxation and without inflammation or hemorrhage; (2) a "hemorrhagic midzone" with contraction band necrosis associated with massive interstitial hemorrhage prominent after reperfusion; and (3) a peripheral, nonhemorrhagic zone characterized by inflammation, phagocytosis and infarct repair. According to this model the myocardium salvaged by reperfusion was 55% after 40 minutes of occlusion, 33% after 3 hours and 16% after 6 hours with 85% of damage after 96 hours of permanent occlusion. Reperfusion after 40 minutes was frequently (41%) associated with ventricular fibrillation (Reimer et al, 1977), with a further note that propranolol protected after temporary occlusion (Reimer et al, 1976) and overestimation of infarct size and understimation of collateral flow were caused by edema, hemorrhage and acute inflammation (Reimer et al, 1977). This model, imitated by innumerable other experiments over the past 20 years, has influenced theoretical and practical approaches to reduce or limit infarct size according to the questionable (Baroldi, 1984) rationale that a small infarct is benign and that around it exists an ischemic border zone at risk of infarction.

Each of the previous points needs discussion. First, acceptance of the "wavefront" concept occurring as it does in the dog with a functioning extramural collateral system is unreliable when its timing is translated to human myocardial infarction. Second, the progression and percentage of salvaged myocardium in that study relates only to the posterior papillary muscle, (see previous discussion) yet the findings are considered a "reliable index of overall left ventricular necrosis". This is not a very convincing index if one thinks of the marked difference in vascularization and function between a papillary muscle and the left ventricular free wall stretched by the interventricular pressure. Third, what is the relation of perfusion to hemorrhagic contraction band necrosis?

The concept of borderline ischemia also seems contradicted by an increase of blood flow (Hood, 1970) and contractility (Goldstein et al, 1972) around an infarct. Furthermore, a wavefront of necrosis, as described experimentally, was never observed in 100 Italian (Baroldi et al, 1974) and 100 Canadian (Silver et al, 1980) fatal acute infarcts and 208 sudden/unexpected deaths (Baroldi et al, 1979). Young infarct necrosis at the periphery of an older one, as a morphologic expression of an expansion of the first damage, was never documented nor hemorrhage within or beyond the infarcted area. It must be stressed that except in unusual circumstances such as wall rupture or following thrombolytic therapy (Fujiwara et al, 1989), the human infarct is never a hemorrhagic lesion during its whole course. Myocardial hemorrhage, which extended even in noninfarcted myocardium, was present in patients who had had fibrinolysis and percutaneous transluminal angioplasty but not in those treated by angioplasty alone. In the latter, plaque fractures and cracks were free of intimal and medial hemorrhage but present in patients with both treatments. The conclusion was that hemorrhage was a consequence of fibrinolytic treatment rather than reperfusion (Waller et al, 1987). On the other hand, the infinite number of procedures to reestablish myocardial blood flow in humans, during the past 40 years, never resulted in fatal malignant arrhythmia affecting half of the cases as occurred in the wavefront experiment. This is a fact which makes legitimate the question whether these procedures really reperfuse a hypothetic chronic ischemic myocardium or their positive effects are due to other mechanisms or to the action of other always associated, drug therapy.

All supporters of the need to reduce infarct size and prevent its expansion should consider that in our selected series of fatal acute infarcts no relation was found between infarct and either death or survival. Also noteworthy is that we observed around the infarct and in other normal zones, a non-hemorrhagic contraction band necrosis interpreted as due to noradrenergic hyperactivity likely triggered by mechanoreceptors (Malliani et al, 1979) or other still undetermined factors, following a loss of function in the infarcted myocardium. This is a kind of adrenergic myotoxic wavefront which fits well with the previous experimental data. Arrhythmogenic complications and/or death are likely related to the degree and maintenance of the latter. Positive technetium-99 stannous pyrophosphate myocardial scintigrams correlated with a complicated postinfarct course and "myocytolytic degeneration" (Buja et al, 1977).

In our experimental infarct studies both CBN and ventricular fibrillation were prevented by beta-blockers (Baroldi et al, 1977) as was myocardial necrosis after transient ischemia (Sommers et al, 1972) or following denervation in permanent coronary occlusion (Jones et al, 1978); lidocaine (Nasser et al, 1980); superoxide dismutase (Przyklenk et al, 1986) or regional preconditioning (Przyklenk et al, 1993). Abnormal myocardial function and flow, in conscious dogs with coronary narrowing, were also prevented by beta-blocking agents (Tomoike et al, 1978). More recently, coronary occlusion in dogs lasting 18, 20, 40 and 60 minutes or after conditioning (ten minutes occlusion + five minutes reflow x four time) produced a progressively increasing extent in the number of foci and myocardial cells with CBN x 100 $mm^2$ with a maximum demonstrated in the "conditioned" group. The necrosis had a similar extent in both "ischemic" and normal myocardium without relation to the blood flow calculated by radioactive microspheres. Both myonecrosis and ventricular fibrillation were prevented by a betablocker (presented for publication). An increase of noradrenaline in the interstitial fluid of an ischemic myocardium has been documented (Lameris et al, 2000). In a practical sense, the measurement of infarct size by serum enzymes or other released substances raises the question whether the latter originate from and in relation to the extent of CBN in which blood flow is maintained or from the infarct necrosis which is sequestered in less than half hour (avascular area) with flow highly reduced, if not abolished, thereby affecting the inflow of therapeutic substances on the infarcted myocardium. Elevation of creatinine kinase and its MB subfraction after bypass surgery (Calif et al, 1998) are likely due to noradrenaline myonecrosis rather than diffuse coronary atherosclerosis (Kini et al, 1999), emphasizing that the concept "a necrosis is a necrosis" contrasts with the need for specific preventive and therapeutic approaches for each form of myocardial injury.

### *Sudden/Unexpected Death*

Amongst 208 sudden/unexpected deaths, 35 had a silent infarct. These are examples of typical silent infarct necrosis always associated with CBN and cardiac arrest likely due to ventricular fibrillation. Several observations support a similar cardiac arrest in the other 173 SD cases in whom the only acute myocardial lesion found was CBN in 72% of cases, without any correlation with the degree of coronary atherosclerosis. Amongst 28 SD cases without or with mild (<50%) coronary stenosis—all cases SD 1st episode—78% had this type of acute myonecrosis.

The reasons why it is possible to indicate that sudden/unexpected death is, in most cases, an event which occurs in the absence of an infarct, are given by studies on resuscitated people and individuals dying suddenly while monitored by a Holter recorder. In two clinical follow-up investigations in a selected population of ischemic heart disease patients successfully defibrillated out-of-hospital, the data were: in one study only 19% of 305 patients developed a transmural myocardial infarction, while 42% had ST and T wave changes and 71 had no appreciable ECG changes. In 38% of cases lactic dehydrogenase isoenzyme was present, and interpreted as secondary to resuscitation maneuvers rather than an index of a small infarct (Cobb et al, 1975, 1980). In another study of 142 victims of out-of-hospital cardiac arrest (all ischemic heart disease patients), tachycardia/ventricular fibrillation was the cause in 95%. Sixty-two of these patients (44%) had had an acute myocardial infarction; 49 (34%) had an ischemic event ("enzymatic ischemia" and/or ST and T changes or left bundle branch block), and 31 (22%) a primary arrhythmic event, (12 with and 19 without ST and T changes (Goldstein et al, 1981,1984).

Another clinical observation deals with people, mainly ischemic heart disease patients, who die suddenly while wearing a Holter recorder. One-hundred-fifty-seven cases of unforeseen sudden death or "ambulatory sudden death" patients with stable health status (cases with an acute myocardial infarction or unstable angina or in a terminal stage were excluded) reported in the literature were reviewed (Bayes de Luna et al 1989, 1990). Most (83%) died from ventricular tachyarrhythmia/ventricular fibrillation. A minority (16%) from bradyarrhythmia/asystole. There was no relation to exercise since sudden demise occurred mainly at rest or when a patient was sleeping.

Three mechanisms of tachyarrhythmia/ventricular fibrillation were identified: (1) ventricular fibrillation preceded by only one premature ventricular contraction or by a very short run of ventricular tachycardia (8%); (2) ventricular tachycardia (more rarely ventricular flutter) that precipitated sudden death usually through ventricular fibrillation, rarely directly or through idioventricular rhythm (62%) and (3) torsades de pointes (13%). The incidence of ST changes, mainly ST depression ≥ 1 mm in patients with ventricular fibrillation, or ventricular tachycardia leading to ventricular fibrillation, was low (13%). In contrast, cardiac pain was present in 33%, dyspnea 26% and the overall incidence of "ischemic" events higher than 70%. These observations are limited to a seriously ill population and the Holter monitor records only one or two leads. The electrocardiographic morphology of sudden death in the general population is still unknown (Bayes de Luna et al, 1990) as is the structural counterpart of different types of cardiac arrest.

The importance of sudden death, epidemic in our society, suggests a survey of findings in the literature on this specific point.

## Sudden Coronary Death in Literature

Any reviewer of the literature on sudden death faces different criteria of selection, dissimilar methods of examination and divergent definitions making an exact comparison of data difficult. For instance, in several studies only cases with at least one coronary stenosis greater than 50% (lumen diameter) or 75% (luminal area) were included (Roberts et al, 1979; Farb et al, 1995). At first glance such selection seems correct in relation to the postulate that sudden death can be coronary only in the presence of a functionally obstructive lesion. However, this type of selection may bias any conclusion. The general impression is a lack of discrimination in these studies between morphologic findings linked with sudden death and those that might be related to: (a) chronic aspects of ischemic heart disease (unexpected versus expected); or (b) acute complications, e.g., coronary thrombus, which may not be responsible for death; or (c) medical procedures such as prolonged therapy and/or resuscitation maneuvers (iatrogenic changes). A similar lack of discrimination exists between initial and terminal morphologic features in relation to survival time. Finally, the frequent lack of control subjects, e.g., accidental death or noncardiac patients in such studies makes unpredictable the pathogenic significance of any lesion found.

With these reservations in mind, we synthesized the main pathological reports related to sudden coronary death from the literature. In referring these data and opinions we have been selective because information may not have been reported in reviewed papers. The aim is to present, avoiding any comment on facts (or fictions?) which allow argument, and offer an objective prelude to a review of the natural history and functional significance of pathologic changes considered the core of various etiopathogenic hypotheses in coronary heart disease.

Amongst 23 studies (Table 38) the man/woman ratio (M/W) ranged from 1 to 16 (mean 6 ± 4). In four reports, where both whites and blacks were included, the M/W ratio was less amongst blacks. The mean age per gender was referred to in only 5 of the 23 papers. In all but one the mean age of women who died suddenly was higher (W 63±6, M 58±4).

### *Coronary Atherosclerosis*

Frequently, amongst reviewed reports, the degree of coronary artery lumen reduction was not cited or was calculated using methods that are difficult to compare. We preferred to distinguish functional and nonfunctional stenoses according to the definition given in papers when such a distinction was clearly outlined. From nine papers the number of SD cases with mild, nonfunctional, stenosis in the whole coronary arterial system was 269 (16%) among the total 1648 (Table 39).

### *Acute Coronary Thrombosis*

The frequency of an associated occlusive coronary artery thrombus ranged from 4% in cases with "instantaneous" death to 82% in sudden death with a survival of 24 hours (Friedman et al, 1973). The average frequency of occlusive coronary thrombus amongst 4524 cases of sudden coronary death reported in Table 40 was 29%. In the 27 papers cited in this Table, only four mentioned the frequency of mural thrombi (Crawford 1961; Newman et al, 1982; Warnes et al, 1984; Davies et al, 1984). Amongst those 310 cases 22% had mural (or intraluminal/intraintimal after Davies) thrombi. Their frequency in these studies ranged from 3 to 35%. There is agreement in the studies that the occlusive coronary thrombus was generally located at a site of a severe "functional" luminal stenosis (Baba et al, 1975; Warnes et al, 1984; Davies et al, 1984).

The frequency of plaque rupture was mentioned in eight studies as follows: 12% amongst 75 cases, (Crawford et al, 1961); 54%, five associated with thrombus, (Friedman et al, 1973); 5% (Liberthson et al, 1974); 31%, 11 with thrombus, (Baba et al, 1975); 10% all with thrombus (Warnes et al, 1984); 4% (Arbustini et al, 1991). In one study (Davies et al, 1984) "plaque fissuring" was present in 103 of 115 vessels showing either mural or occlusive thrombi.

### *Acute Myocardial Necrosis*

The frequency of acute myocardial infarct necrosis was referred to in 21 studies (Table 39). Infarct size was not considered but in seven studies the age of the infarct was estimated histologically (Bedford, 1933; Levy, 1936; Jorgensen et al, 1968; Scott et al, 1972; Liberthson et al,1974; Baba et al, 1975; Haerem, 1975; Reichenbach et al, 1977). Of these 294 infarcts, 41% were estimated to be older than 24 hours. The frequency of an acute occlusive thrombus associated with an infarct ranged from 22-68% (mean 43%) in studies in which this variable was reported.

### *Type of Coronary Distribution*

Preponderance of one coronary artery has been considered a pathogenic factor in ischemic heart disease (Schlesinger, 1940). In some studies of sudden coronary death the predominant artery was documented. However, no relation between the type of coronary distribution and sudden death was specified. In general, the left anterior descending branch was the artery more often involved by atherosclerosis and thrombosis in cases of sudden death ("artery of occlusion and sudden death" Barnes et al, 1932). A fact confirmed by angiography in patients with ischemic heart disease, who subsequently died suddenly (Vlay et al, 1993).

### *Collaterals*

Only one pathological study investigated intercoronary collaterals in sudden and unexpected death cases (Spain et al, 1963). That study, by postmortem intracoronary injection of calibrated (40 to 75 μm) plastic beads followed by injection of a warmed suspension of barium in gelatin and subsequent histologic examination of the myocardium, indicated the presence of intercoronary collaterals diagnosed by movement of plastic beads from one coronary artery to another; increased X-ray vascularity and histologic evidence of giant capillary-like vessels in 10 of 13 SD cases with healed myocardial infarcts and in only 1 of 16 SD cases with advanced coronary atherosclerosis but without infarct. No anastomotic channels were seen in 76 normal subjects with different degrees of coronary atherosclerosis who died by accident. "The absence of collaterals larger than 40 mm might explain sudden death following an acute ischemic attack..." and an infarct was believed "...a prerequisite for the development of functionally significant intercoronary anastomoses" (Spain, 1963).

### *Heart Weight*

In ten studies (Table 41) the frequency of heart weight following our definition of pathological heart weight (≥ 500 g) could be calculated. It was 46% of 1279 cases. It is interesting to note that amongst 115 young soldiers who died suddenly (Moritz, 1946) none had a pathological heart weight. In reviewing data in these ten studies it was impossible to correlate heart weight and extensive myocardial fibrosis.

### *Morphologic Variables of the Atherosclerotic Plaque*

In the literature a systematic correlation between different physical and morphologic variables in coronary arteries and postmortem findings in the various clinical patterns of ischemic heart disease was lacking. It was shown that in acute infarction, sudden death and unstable angina patients the mean percent of dense fibrous tissue, calcific deposits, and pultaceous debris increases with increasing degrees of luminal narrowing while the mean percent of cellular fibrous tissue decreases (Kragel et al, 1989). In other studies active inflammation was significantly related to a reduction of the arterial lumen and a cardiac cause of death (Cliff et al 1988). A high frequency of T and B lymphocytes was observed in 11 atherosclerotic plaques of patients with unstable angina (100% in the adventitial and 82% in the intima), in 45 plaques of patients with acute myocardial infarction (87% and 91%), in 18 plaques of sudden death cases (72% and 83%) and in 15 plaques of patients dying of noncardiac disease (73% and 65%). In the same study the amount of preserved media, expressed in percentage of the media in histologic sections, was calculated at 10 different levels in each of 90 plaques. The preserved media was 35±9% when a thrombus was present and 29±10% when absent. This difference was not statistically significant (Arbustini, 1991). In particular, intimal hemorrhage alone or associated with plaque rupture and/or fissuring had a low frequency (Table 42).

### *Conduction System*

The complexity of studying the conduction system by serial or semiserial section study limited its examination to a few investigations. One is represented by 30 clinico-pathologic papers ("De Subitaneis Mortibus") published in *Circulation* from August 1975 to June 1978 (James et al, 1973-1978; Brechenmacher et al, 1976,1977). Amongst 77 sudden death cases reported, 60 were associated with different underlying diseases, seven with conduction disturbances (varying degrees of heart block, paroxysmal atrial fibrillation, premature ventricular beats, etc) and ten had no history of associated disease or precursor episodes. The pathologic findings observed in these 77 cases ranged from benign tumors, fibroumuscular medial hyperplasia, focal neuritis, replacement of adipose tissue, etc. This series of case reports is a miscellanea, while the need is for systemic studies of those diseases more prone to sudden death with findings checked in a normal population and other diseases in which sudden death is rare or absent.

Another study (Lie, 1975) examined the conduction system in 35 men and 14 women; (mean age 57 and 62 respectively) with ischemic heart disease. The subjects, 39 with, and 10 without, histologically demonstrated myocardial infarct died suddenly within six hours of the onset of acute symptoms. Most hearts had severe (>75% cross sectional area) coronary atherosclerosis, acute myocardial ischemia expressed by myofibrillar degeneration and cardiomegaly (> 350 g). In both groups 40% of cases had a healed myocardial infarction while a recent coronary thrombus was demonstrated in 18% of infarct and 20% in noninfarct cases. Findings in the conduction system of all cases were: intimal thickening with severe luminal stenosis of the sinoatrial artery in 26% or atrioventricular node artery in 52%, destruction of the atrioventricular node in two instances with massive interventricular septal infarction, fibrosis and/or fatty replacement of sinus node (9 cases), atrioventricular node (2), His bundle (22), right (4) or left (14) or both bundle branches (7). Dysplasia of the atrioventricular node artery was present in 44% of 27 sudden death cases vs 6% of 17 controls (Burke et al, 1993).

## *Sudden Coronary Death and Exercise*

The risk of strenuous exercise in coronary heart disease is controversial (Gibbons et al, 1980). A review of studies where "physical effort" was investigated (2324 cases) shows that sudden death occurs mainly while a patient is at rest or sleeping or during minimal activity (69%). The remaining 707 subjects were at work.

In this regard, pathologic findings observed in selected groups such as young athletes and "joggers" who die suddenly must be mentioned. In 78 "athletes" reported in the literature (Heath, 1969; Opie, 1975; Noakes et al, 1979; Maron et al, 1980; Morales et al, 1980; Waller et al, 1980; Tsung et al, 1982; Virmani et al, 1982; Voigt et al, 1982; Thiene et al, 1983, 1985) the pathology found at autopsy was mainly cardiac (Table 43) and generally related to chronic disease. Only in three cases, one with pulmonary thromboembolism and, two with a ruptured aorta was an acute event possibly linked with the death. The age distribution of these 78 cases (Table 44) shows that coronary atherosclerosis was estimated as a cause of death in 3% of subjects under the age of 20 years, 40% between 20-29, 67% between 30-39 and 100% after 40 years. Similarly in 36 joggers (Morales et al, 1980; Virmani et al, 1982; Thiene et al, 1985) who died suddenly (Table 45) coronary atherosclerosis was the estimated cause of sudden death in 92%.

In most sudden death cases reported in the literature, the referred causes were an old lesion preceding death long before and despite, as in athletes, repetitive psychological and physical efforts. The need is to recognize how many sudden deaths are sudden cardiac death as documented by contraction band necrosis; avoiding any insistence that the latter or other inconsistent changes indicate ischemia as the cause of the sudden demise. Some of these cases seem appropriate to discuss the morphofunctional significance of pathologic findings. For instance, that of a 12-year-old girl with an enormous fibroma of the left ventricle (Heath, 1969) who suddenly died following a swimming race, without any history of cardiac symptoms or signs despite her active athletic life; or the two following cases of marathon runners (Noakes et al, 1979):

1. a 44-year-old man who had been running for 14 months completing eight standard 42 km marathons, a 56 km race in five hours and 59 minutes, 90 km marathon in 10 hours and 10 minutes. In training he ran from 48 to 80 km per week. He had no history of ischemic heart disease and smoking. Following a 50 km race (4 hours and 59 minutes) and, one month later, a standard 42-km marathon (4 hours and 2 minutes) he visited a general practitioner complaining of a "nonspecific lack of energy". There were no symptoms or signs of ischemic heart disease and all tests were negative. No electrocardiogram was performed. Shortly afterward he competed in a 24 km race. At the 19 km mark, while running without distress he stopped to adjust a loose shoelace. Bending down he suddenly lost consciousness and died instantly. At autopsy the main findings were: heart weight 357 g; extensive old antero-septal scar (old "silent" infarct); grade 4 (75-100%) lumen reduction in the first part (5 mm) of the

left anterior descending branch and associated total occlusion by organized and recanalized thrombus. The other severe (75-100%) coronary lesion found was 3.5 cm from the origin of the left circumflex branch. There was no evidence of a fresh infarct; only foci of "contraction band necrosis" were present in both ventricles. The conduction system was normal.

2. 41-year-old man, who had been marathon runner for two years, suffered an acute myocardial infarction. He had a complete angiographic occlusion of the left circumflex branch, 50% narrowing of the proximal tract of the right coronary artery and minor luminal irregularities of the left anterior descending vessel (March 1976). Ignoring medical advice and after appropriate training, approximately one year after the infarct, he completed a 50 km race in five hours and 36 minutes. Two weeks later he ran a 42 km marathon in four hours 45 minutes, followed by three additional marathons in the following three months. From the time of infarction to his last race (28 months) he ran 3624 km. Pain in the chest, jaw or left arm occurred several times during training runs but he did not seek medical advice. Subsequently, he was readmitted to hospital for two episodes of unstable angina. Angiography confirmed the complete occlusion of the left circumflex and revealed total occlusion of the right coronary artery and 80% stenosis of the left anterior descending branch (October 1978). While awaiting coronary by-pass surgery, he began to have severe chest pain, with electrocardiographic signs suggestive of an acute myocardial infarction, i.e., acute antero-lateral ST-segment elevation with increasing size of Q waves in $V_5$ and $V_{6)}$. Despite therapy the heart stopped within 30 minutes and resuscitation failed. There was no evidence of enzyme changes. At autopsy the heart (349 g) showed: 75-100% obstruction in the first tracts of the left anterior descending, left circumflex branches and the right coronary artery, with superimposed fresh thrombus in the left descending, organized thrombus of the left circumflex and organizing thrombus of the right coronary artery. A healed infarct of the left ventricular inferior wall, its appearance in keeping with the infarct suffered two years before, was documented. An acute myocardial infarction was not demonstrable. Foci of contraction band necrosis were observed. The conduction system was normal.

Finally three sudden death cases with a "mural" left anterior descending branch and strenuous exercise (Morales et al 1980) deserve attention:

1. For many years a 54-year-old man had episodic left precordial pain radiating to his back and induced especially by emotional stress. ECG demonstrated subendocardial ischemia with a strongly positive exercise stress test. Coronary angiogram showed a "milking effect" (Noble et al, 1976) in the proximal portion of the left anterior descending vessel. The ventricular function was normal. He died suddenly while jogging. Pathological findings were limited to the heart. It weighed 440 g. A 2.5 x 1.3 cm scar with intermixed brown myocardium involved the most anterior portion of the ventricular septum and adjacent left anterior ventricular wall from the apex to within 2.5 cm of the base. Coronary arteries showed only two plaques with 50% lumen reduction of the posterior right coronary artery and first segment of the left descending branch. Two centimeters from its origin the left circumflex was covered for 4.5 cm by a loop of atrial muscle, while the left anterior descending immediately after the atherosclerotic plaque dipped into the myocardium for 3 cm. Histologically, patchy irregular areas of healing necrosis alternated with areas of normal or injured ("muscle with irregular outlines, myocytolysis, fragmentation") myocardium in the anterior septum and left ventricular wall. The atrioventricular nodal branch was stenosed by concentric intimal hyperplasia.
2. A 34-year-old man without a medical history died suddenly while jogging. The heart (460 g) showed a "mural" disposition of the left anterior descending coronary branch for 2 cm, 3.5 cm from its origin. No atherosclerosis was found in the coronary arteries. Grossly and histologically myocardial changes were similar to that described in case 1.

3. A 17-year-old woman, who while swimming in a pool and after completing approximately 10 laps, helped herself from the pool but immediately thereafter became unconscious, with no vital signs. She died in hospital 14 hours after resuscitation attempts. Autopsy findings were marked congestion and edema of the lungs. The heart (260 g) was free of coronary lesions. The right coronary artery was overbridged for 3 cm in its posterior tract. At its origin the left anterior descending branch dipped into the myocardium for 2 cm and for 1.5 cm in its terminal segment. The endocardial half of the entire left ventricle was dark red with a histological pattern of hemorrhagic necrosis where necrotic myocardial fibers (often with contraction bands) were spread apart by interstitial hemorrhage and an early polymorphonuclear leucocytic infiltrate. No significant myocardial fibrosis was present.

On the significance of mural coronary arteries, one notes that ischemic heart disease patients successfully treated by surgical coronary debridging have been reported (Faruqui et al, 1978).

## Interpretation of Coronary Syndromes

The presently accepted viewpoint on the acute coronary syndromes—infarct, sudden death, unstable angina—is that they have a common etiopathogenesis related to plaque rupture/thrombosis and their complications. The following synopsis, synthesizing the dogmatic and antidogmatic interpretation of main facts pertinent to the natural history of coronary disease, may help our discussion.

***Synopsis: Arguments in favor of or against a cause-effect relationship between ischemia and coronary syndromes***

| Dogma | Antidogma |
|---|---|
| 1. Experimental occlusion of a normal coronary artery results in infarct and possible ventricular fibrillation. In each species the infarct size is stable for that animal. | In human CHD eventual occlusion occurs related to an old severe stenosis bypassed by functioning collaterals. Myoinfarct size ranges from less than 10% to >50% without relation to the territory supplied by the related coronary artery. |
| 2. Progressive expansion of original experimental infarct caused by ischemia and achieved by contraction band necrosis (CBN) plus hemorrhage after reperfusion or longlasting occlusion. Ischemic "wavefront phenomenon" often associated with ventricular fibrillation. | Extensive, nonhemorrhagic contraction band necrosis (CBN) is found surrounding infarct necrosis and in normal areas elsewhere. After reperfusion in humans hemorrhagic CBN and ventricular fibrillation are rare. |
| 3. Myonecrosis is a unique entity and "infarct" interpreted only as an ischemic lesion. No consideration that different forms of myonecrosis with their own etiopathogenesis, exist. | necrosis is associated with nonischemic CBN (or coagulative myocytolysis) due to catecholamine myotoxicity or nonischemic colliquative myocytolysis. |
| 4. Size and expansion of the infarct is the cause of death. | Size of infarct necrosis is not related to cardiac arrest. Complications and death are mainly due to ventricular fibrillation (CBN) or congestive heart failure (colliquative myocytolysis). |

***continued on next page***

### *Synopsis: Continued*

| Dogma | Antidogma |
|---|---|
| 5. Coronary collaterals have no functional meaning in protecting human ischemic myocardium. | First episode of CHD in apparently normal subjects occurs in the presence of single or multiple old coronary stenoses bypassed by enlarged collaterals as shown by tridimensional coronary plastic casts. Similarly, severe obstructive coronary atherosclerosis without clinical and pathological evidence of ischemia is frequent in population of normal people and noncardiac patients. Coronary occlusion after a few days of an experimental critical stenosis does not produce an infarct or dysfunction as a result of a dramatic increase of collaterals. |
| 6. Atherosclerotic plaque rupture is a primary event and induces occlusive or mural thrombus. Thrombus gives rise to myocardial microemboli which in turn causes acute coronary syndromes. | Plaque rupture/thrombosis, when present, are events secondary to changes at the plaque level (collateral flow) following intramyocardial peripheral resistance. Atheromatous intramyocardial emboli should be a frequent finding. One embolus only in more than 500 CHD hearts systematically studied. |
| 7. Plaque rupture is proved by material injected in the coronary arteries which enters subendothelium via endothelial fissuration. | Material injected in a coronary artery reaches the subendothelium via collateral plaque vascularization. |
| 8. In two studies, patients with unstable angina who die suddenly within 6 hours have fibrin-platelet thrombi or emboli in small intramyocardial and "microinfarcts". | Most of these patients reported in the literature had a histologically documented acute infarct. In this condition secondary fibrin-platelet thrombi are part of infarct necrosis. "Microinfarcts" are foci of catecholamine CBN. |
| 9. Angiographic demonstration in vivo within 1-4 hours of coronary occlusion in 87% of AMI patients. | The unique AMI case monitored angiographically showed a secondary pseudocclusion starting distally (intramural resistance) and ascending at the origin (not at plaque level) of the related artery.The question is how many of 87% angiographic occlusions were pseudocclusions. |
| 10. Recovery of a "layered" thrombus in AMI patients at emergency bypass surgery. | The "layered" blood is not a thrombus but a coagulum which forms in progressive blood flow reduction. |
| 11. Occlusive thrombosis/infarct due to rupture in small plaques undetectable angiographically in vivo. In angiograms long before AMI showed small plaque in the artery supplying the zone of future infarction. Since other vessels had the unchanged small plaques seen previously, deduction was that similarly the infarct related artery maintained the same small plaque. | If it exists it must be proved histologically. Not observed postmortem in all pathological studies. This hypothesis does not consider plaque progression due to hemodynamic wall stresses secondary to myocardial-related asynergy which precedes infarction. |

***continued on next page***

***Synopsis: Continued***

| Dogma | Antidogma |
|---|---|
| 12. Congestive heart failure due to myocardial fibrosis following ischemia or viral myocarditis or thickening collagen fibrillar matrix, secondary hypertrophy, apoptosis, myocardial slippage, myocellular neogenesis or adrenergic overactivity. | In hearts excised at transplantation from patients with identical clinical parameters of CHF despite different diseases (CHD, dilated cardiomyopathy, valvulopathy) quantification of lesions does not support any proposed theories, but adrenergic hyperactivity. |
| 13. Atherosclerotic plaque begins as fatty streak by transendothelial insudation of lipoprotein-cholesterol and inflammation, as shown by experimental hypercholesterol diet and in hypercholesterolemic patients (hypercholesterol plaque). | In CHD patients and normal subjects the coronary atherosclerotic plaque starts as nodular smooth muscle cell hyperplasia, elastosis and fibrous/repair. Lipoprotein/cholesterol+macrophages collect in a proteoglycan pool, deep to the fibrous cap (myohyperplastic plaque). This is a pattern totally different from the hypercholesterol plaque. |
| 14. Divergencies of plaque variables in different patterns of CHD and controls were never systematically investigated. | Frequency and extent of coronary plaque variables differ in various CHD groups and controls mainly in relation to hemodynamic changes within the plaque secondary to intramural resistance, being essential to discriminate primary and nonprimary changes. |
| 15. Plaque progression due to further lipo-protein material, platelet-fibrin thrombi and inflammation. | Coronary plaque progression proceeds by recurring myohyperplasia+fibrosis, proteoglycan accumulation and atheroma and/or calcification. Progression is mainly influenced by hemodynamic stress pressures related to subtending asynergic myocardium. Aggravation of the latter explains both a more rapid progression of the plaque and restenosis after angioplasty rather than does thrombosis. |
| 16. Atherosclerotic plaque inflammation is due to macrophages, T lymphocytes. | Macrophagic reaction is a repair and not an inflammatory process. T lymphocytes and plasma cells, particularly around nerves of coronary and aortic medial appear following proteoglycan accumulation in the deep intima (medial neuritis) and not in relation to lipoprotein/cholesterol. |
| 17. Cause of sudden death generally related to chronic lesions and CBN, when reported, is interpreted as an ischemic lesion. | Sudden/unexpected coronary death is a nonischemic event triggered by zonal or focal adrenergic stress +CBN+ ventricular fibrillation. Cases with histologically demonstrated infarct should be better defined as rapid death associated with an infarct or sudden expected death. |
| 18 Experimental reperfusion necrosis (RN), i.e., CBN+massive interstitial hemorrhage is considered a part of infarct necrosis. | In CHD, even if associated with infarct necrosis, CBN is never hemorrhagic. RN in man is a rare finding following heart surgery or resuscitation. RN=CBN+capillary wall damage. |

## Personal Interpretation Based on Our Morphologic Experience

In describing the pathological background of the coronary syndromes, each pattern has (1) its own morphologic and clinical physiognomy unrelated to atherosclerotic plaque rupture/thrombosis and (2) pathognomonic early and late findings with a different pathogenesis and still undetermined etiology.

### *Pathogenesis*

#### Acute Myocardial Infarction

We believe that this acute syndrome begins with hypo-akinetic myocardium in relaxation, a fact recognized as first ischemic event long ago (Nesto et al, 1987; Taki et al, 1987). The affected myocardium becomes acutely stretched by the pulsating intraventricular pressure until it blocks nutrient flow as a result of intramural extravascular compression brought about by a dyskinetic bulging (transmural) or compressive (subendocardial or internal infarcts) effect suggested by early elongation of sarcomeres and nuclei and by the "avascular area" .One notes that (a) in an artificial system flow within a normal lumen stops (extramural angiographic pseudoclusion) when there is a maximal increase of peripheral resistance (Gregg, 1950); (b) the mechanism of blockage of nutrient flow leading to infarction is not due to a primary "hydraulic" occlusion of the intramural vessels; and (c) the frequency of an occlusive thrombus, found in about half of the cases, relates to collateral flow redistribution, particularly at the plaque level with all secondary changes, in proportion to the extent of infarct size. Experimentally, if the cardiac wall of a sheep's heart was "protected" by a Marlex mesh positioned before a coronary occlusion, there was no massive scar, aneurysm or cavity dilatation and minor hemodynamic and contractile dysfunction in contrast to controls (Kelley et al, 1999). This is a reliable experimental model which may be useful to investigate and quantify all the changes of myocardial infarction, including CBN wavefront, in relation to dyskinetic bulging.

We found that the size of infarct necrosis, in the absence of wall rupture plus tamponade, is not the cause of complications and death. All infarcts, independent of their size and associated degree of coronary obstruction, are surrounded by an extensive zone of nonhemorrhagic contraction band necrosis, foci of which are also found in normal myocardium elsewhere. In reality, the wavefront which experimentalists describe and clinicians now accept blindly, is an expanding catecholamine-induced necrosis. The beneficial clinical effect of betablocker agents is to prevent adrenergic overstimulation and this form of necrosis which is linked in turn with malignant arrhythmias/ventricular fibrillation (see in Chapter 6).

The other histologic change observed in about half of acute infarcts is colliquative myocytolysis found in the layer of myocardium, preserved by infarct necrosis, in subendocardium and around functioning vessels at the periphery of an infarct. This histologic hallmark indicates the other nonischemic complication of myocardial infarct (see below).

#### Sudden/Unexpected Coronary Death

Sudden/unexpected coronary death is not necessarily synonymous with the development of a myocardial infarction as demonstrated in most resuscitated people. Apart from the cases of silent infarct—with findings similar to nonsilent infarcts (Table 46), the only significant difference being the lack of subjective symptoms—sudden death seems linked to a primary adrenergic overactivity shown by the association between CBN and ventricular fibrillation, in the absence of congestive heart failure. It is important to recall that sympathetic dysfunction must be regional since no arrhythmias were demonstrated experimentally by systemic intravenous infusion of catecholamines (Todd et al, 1985b) and sudden cardiac death is not reported in the clinical course of human pheochromocytoma; while severe malignant arrhythmias occurred in a few dogs with norepinephrine infusion in one coronary artery (unpublished data). An increase of interstitial noradrenaline has been documented in ischemic myocardium (Lameris et al, 2000).

### Angina Pectoris

Angina, whether unstable or stable, is an acute clinical syndrome accessible to a pathologist only when it evolves into an infarct or sudden death. Nevertheless, it presents the same background in terms of coronary atherosclerosis as shown angiographically, associated or not with zonal myocardial hypokenesis. We note that in the reported case of unstable angina cineangiographically monitored and mentioned above there were no images of a "thrombotic" plaque nor evidence of intramural lesions in the excised heart at transplantation. There is no way to prove morphologically that CBN forms at the time of the clinical crisis. However, from all data observed in the two other syndromes it is likely that the adrenergic system has a role also in this pattern.

### Chronic Coronary Heart Disease—Congestive Heart Failure (CHF)

Chronic coronary heart disease is often erroneously defined as ischemic cardiomyopathy related to congestive heart failure (CHF). In a recent "consensus" (Cohn et al, 2000), the latter—and not only following CHD—has been considered secondary to absolute or relative ischemia with subsequent remodeling of cardiac shape by enlarged volume and increased mass, thinning of heart wall and lengthening of cardiomyocytes, continuous infarct expansion, scar formation, apoptosis, hypertrophy, slippage of myocardial cells and increased interstitial fibrosis. A galaxy of events in which proinflammatory cytokines and tumor necrosis factor (Feldman et al, 2000) and neurohormonal activation play a fundamental role (Cohn et al, 2000).

In our study of 63 CHD, 63 dilated cardiomyopathy and 18 valvulopathy hearts excised at transplantation from patients all with an identical clinical pattern of irreversible congestive heart failure, we could not confirm (Baroldi et al, 1998) many of the changes mentioned by Cohn et al (2000). First, acute silent infarct necrosis was found in only a few coronary heart disease cases without clinical evidence of ischemia. Then the extent of myocardial fibrosis was deemed not great enough to explain dysfunction even in those hearts from CHD patients. The "fibrotic index", i.e., the extent of fibrotic area in percent of the total histological area, showed that a histologically viable myocardium was 83% in CHD, 98% in dilated cardiomyopathy and 94% in valvulopathy. Furthermore, there were no objective findings to prove that a lymphocytic myocarditis caused fibrosis and failure. The number of lymphocytic foci x $mm^2$ and number of lymphocytes per focus—particularly in dilated cardiomyopathy—were less than in controls. Furthermore, the undulate or wavy collagen fibers observed in fibrotic areas (Fig. 18) can be considered an adaptation by the proliferated collagen within a beating myocardium; having in mind that this interstitial fibrosis may be due to both a repair process of microfocal CBN and collagen matrix proliferation (Rossi et al, 1998) without any constrictive effect. The blood flow reduction recording in patients—with or without CHD—with congestive heart failure was not related to myocardial fibrosis (Parodi et al, 1993; De Maria et al, 1996) being likely due to diminished contractility with reduced metabolic demand. Finally in CHF there is no clinical documentation of ischemia because of lack of energy deficiency (lactate production, coronary sinus oxygen reduction and pH) as well pertinent symptoms (Pool Wilson, 1993).

The weight/size paradox pertains to CHF independent of its cause. It means a contradictory increase of heart weight despite thinning of cardiac wall and a normal diameter of myocardial cells; the latter being interpreted as a stretching of myocells and their neogenesis and slippage. In fact, stretching of myocells does not exist by histology and electron microscopy (Fig. 24); keeping in mind that CHF may occur without dilatation of the heart chambers. Similarly, myocellular replication by longitudinal cleavage of the hypertrophic myocells (Linzbach

1947, 1960) has never been proved; or by newly formed myocardial cells as suggested by mitotic myocytes found in failing hearts (Beltrami et al, 1994) and at the border zone, and in normal myocardium in acute human infarct (Beltrami et al, 2001) needs investigation. In the latter condition the number of mitotic myocells was 4% at the infarct border and 1% in normal myocardium, being absent in control hearts. The myocyte mitotic index, i.e., "the ratio of the number of nuclei undergoing mitosis to the number not undergoing mitosis" was very low (0.015) and the claim that, "if sustained, mitosis could result in the formation of 100 g of myocardium in less than three months" is an example of a theoretical calculation out of the reality. The belief is that organs, like brain and heart, are formed by parenchymal specific elements, namely neurons and myocardial cells that, in the adult, are unable to reproduce. A dogma already challanged long ago by replication of these elements in culture. Consequently, what makes a distinction between adult organs capable or not of replication is the possibility that newly formed elements must integrate in the complex functional architecture of a tissue. It is difficult to prove that a new neuron or myocardial cell may become a part of tissue function when the latter needs complex connections betweeen all the elements. As far as the myocardium is concerned, any newly formed myocell must join others by intercalated disc plus intermyocellular myofibrillar bridges and assume a registered order in parallel disposition for pump function. The demonstration of mitotic myocells, per se, does not necessarily mean that these cells will be able to reconstruct a functioning myocardium. The fact remains that even for microfoci of myocell necrosis involving relatively few elements the repair is by scar despite the myocellular potential for replication. The question if the mitotic elements are resident cardiomyocytes or derived from stem cells is a biological problem which at present has little if any clinical application; therapy promoting myocellular mitosis being still unpredictable and unreliable. In our experience, a unique focus of replicated myocells was observed in an endomyocardial biopsy at a previous site of sampling. The new, small, disarrayed myocells were present in atrio-ventricular node-like tissue and were surrounded by normal myocardial cells (Fig. 24); a pattern never seen in any examined hearts. On the other hand, an increased number of myocardial nuclei in a cell by amitotic division but without myocellular neogenesis may suggest myocellular replication. However, amitosis occurs in normal hearts with an increased frequency in hypertrophic and atrophic hearts (Baroldi et al, 1967). Finally, the idea of a slippage of myocardial cells, i.e., their interpenetration resulting in thinning and elongation of cardiac wall (Linzbach, 1947; Weber 1989; Beltrami et al, 1994; Cohn et al, 2000), conflicts with several facts. First, how does hypertrophic myocell become a normal sized cell by slippage while the hypertrophying stimulus persists? Second, how can slippage occur if myocardial cells are strictly interconnected by numerous myobridges and collagen network? Third, how can slippage occur without destroying all interstitial structures, i.e., vessels, lymphatics and nerves, the disappearance of which should produce severe tissue damage, when a wall thickness of 3 cm is reduced to 1.5 cm (Fig. 24)?

A last comment relates to apoptosis causing CHF. Since apoptosis means a losts of cells without fibrotic repair (Majno et al, 1995), hearts dying from failure due to apoptosis should have a normal, if not, lower weight and not an increased one. Having in mind that an atrophic heart rarely becomes an insufficient pump (Hellerstein et al, 1950); atrophy being likely a readaption of the myocardial mass to a reduced functional demand in a cachectic body. In reality, TUNEL techniques detect DNA ending exposure. This may happen anytime fragmentation of a myocardial cell occurs as in contraction band necrosis. In turn, this questions the specificity of TUNEL technique since it may express a repair process in any necrosis (Kanoh et al, 1999); caution is needed when interpreting TUNEL positivity (Jerome et al, 2000). Furthermore, in congestive failing hearts nuclear changes typical for apoptosis were not observed in one ultrastructural study (William et al, 1999) or seen with a frequency ranging from 0.06% to 0.41%, apoptotic nuclei being defined as "chromatin margination, condensation, clumping and discontinuity of nucleare membrane" (Guerra et al, 1999). The pathognomonic, specific change, i.e., apoptotic bodies, as expression of nuclear fragmentation was not mentioned. It is an easily detectable finding histologically but was never seen in our material.

Our CHF study (see above) confirmed two important points. First the existence in the natural history of CHF, independently from the underlying disease, of an adrenergic hyperactivity as shown by the high frequency of any stage of catecholamine myonecrosis, despite its minimal extent. The second morphological finding is given by colliquative myocytolysis, i.e., progressive vacuolization by disappearance of the myofibrils without any cellular reaction, as a hallmark of CHF. Obviously, vacuolization of the myocardial cell may occur in other rare conditions for example storage (glycogen, etc) or parasistic diseases or can be an artifact. However it is mainly interpreted as the result of a reduced blood flow perfusion. Already we stressed the absence of ischemia in CHF. On the other hand, vacuolization is never present in human infarct necrosis, in CBN and never reported in any experimental model of coronary ischemia, both with permanent or temporary occlusion, particularly in the prenecrotic stage. In our cases the extent of colliquative myocytolysis did not correlate with the amount of flow (Baroldi et al 1998; Parodi et al, 1993; De Maria et al, 1996). It must be stressed that the lack of correlation between clinical indices and extent of colliquative myocytolysis indicates the latter as a secondary phenomenon and not a cause of myocardial insufficiency.

**Beneficial Therapeutic Effects in CHD**

One must be cautious of the positive results reported from various "reperfusing" clinical/surgical approaches. They tend to maintain faith in the supremacy of ischemia in CHD. The point is whether these approaches work as they are thought. For instance, streptokinase also improves myocardial contraction in an ischemic isolated heart and therefore in the absence of platelets and fibrin (Fung et al, 1984) while fibrinolysis is stimulated by many factors, e.g., tissue necrosis, catecholamines etc. Several other benefits from different procedures adopted clinically with different intentions can be cited. For example, the assumption that cholesterol lowering therapy, beta-blocking agents and possibly angiotensin-converting enzyme inhibitors and antioxidants reduce plaque rupture (McIsaac, 1993), when the procedures may work at other levels. Furthermore, caution is needed any time an attractive new therapy is proposed. The vascularization obtained by injecting DNA expressing vascular endothelial growth factor in the endocardium at the site of an infarct in rats (Schwarz et al, 2000) seems more neovascularization of an intracavitay thrombus than of the infarcted wall. Similarly, the clinical benefits—improved quality of life, excercise tolerance, increased wall thickening—in CHD patients treated by intracoronary basic fibroblast growth factor (Lahan et al, 2000) badly needs documentation of new vessel formation. Furthermore, the decreased apoptosis, long term salvage and survival of the viable myocardium, reduction in collagen deposition and sustained improvement of cardiac function obtained by human bone marrow-derived angioblasts in experimental infarct of mice have been interpreted as due to proliferation of preexisting vasculature (angiogenesis) or new vessel formation within the infarcted myocardium (Kocher et al, 2001). However, in this study neovascularity has been deduced by endothelial cell density. The latter can be due to angiohyperplasia (see above) rather than new vessel growth. It should be kept in mind that many surgical procedures where no revascularization, or reperfusion, was never documented, abolished chest pain and produced beneficial effects in patients; and a similar outcome was observed following aggressive and nonaggressive therapy (Mark et al, 1994) and between angioplasty and intravenous thrombolysis (Dauchin et al, 1999).

## Etiology

The cause of coronary syndrome is still a matter of investigation, seeking well-grounded hypotheses since the "hydraulic" one is untenable. The question, therefore, is what promotes (1) the initial focal hypokinesis in myocardial infarction/angina and its relation with chest pain or equivalents, e.g., dyspnea, fatigue, etc. (2) primary ventricular fibrillation in sudden coronary death in the absence of an infarct; and (3) nonischemic myocardial dysfunction in congestive heart failure.

With this background let us to explore the relationship between coronary heart disease and the nervous system.

# Chapter 6

# Adrenergic Stress

For millenia mors subita (sudden death) was more a voodoo-religious concept than a bio-pathologic problem. Cases are reported episodically in ancient writings as that of Phidippides, cited above, who died suddenly crying to the Athenians "Nike" (victory) against the Persian, having run 22 mi. 1,470 yd from Marathon to Athens. However, the association between sudden death and heart disease came with Lancisi's publication (1745), the first scientific report based on autopsy findings. Subsequently, sudden death was recognized as part of ischemic heart disease. Do similarities exist between the 18th Century Romans and modern societies? At best, relationships seem tenuous. Modern societies are populous, rich and technologically advanced while Rome in 1700 was at a peak of decadence and poverty and had a small, technologically unsophisticated population. A possible common denominator of the two, apparently opposite, social patterns might be mental depression consequent to loss of hope. Yet, both societies, faced tension, stresses and depression. Eliot (1994) reported an epidemic of sudden death amongst National Aeronautics and Space Administration employees when, in 1968 that agency faced budget cuts with the loss of highly technical jobs not readily available in other industries. This caused environmental instability. Other similar examples were reported to show the relationship between depression and socio-economic factors (Salomon et al, 2000). In the meantime, atherosclerosis, per se, is an old human disease being present in Egyptian mummies of high social class (Ruffer, 1911). However, the deluge of coronary heart disease cases started only when society became affluent in a highly unstable economic and technologic enviroment characterized by many stimuli and related tensions.

In the course of evolution mankind has always paid a high price to survive. Infectious diseases and malnutrition were, and in many areas of the world still are, the main causes of morbidity and death. Where an agricultural life-style has been radically changed to a complex industrial/consumer type system, infectious diseases and malnutrition have been replaced as the major causes of morbidity and death by the "heart attack". The malignity of coronary or ischemic heart disease is indicated by it being the prime cause of morbidity and mortality in Western countries. Also, it generally involves the more competent and competitive people at the time of their highest productivity ("society has need of such men, produces them destroys them" Morris, 1969). For instance, in 1946, when coronary heart disease was the prime killer in United States, in Italy the autopsy population was mainly patients with fatal infectious diseases, particularly tuberculosis; a myocardial infarct was a curiosity. In only 10 years, following an extraordinary industrial/economic boom CHD was as much of a problem amongst Italians. At present in Italy, of approximately 500,000 deaths per year, 200,000 are of cardiovascular origin, mostly due to ischemic heart disease and of the latter 100,000 (about half sudden deaths) occur in people between 40 and 65 years of age. In the United States 500,000-600,000 sudden natural deaths have been estimated to occur each year, most of which (about 90%) are related to ischemic heart disease (Lie, 1991).

In the past and when dealing with infectious epidemics in the dark ages, the first defense was to identify sources of contagion and isolate them. The main risk factors were contact with sick people or infected water, food etc. The main preventive measure was the lazarette, because of limited therapeutic means. This ghetto prevention reduced, in someway, the genius morbi. True preventive and therapeutic measures began only when the etiology of an infectious disease was discovered and immunization or anti-agent medicaments became available. Despite much knowledge, some of which may prove to be dogma, we are still in the dark ages as far as the cause of coronary heart disease is concerned. Its pathogenesis is not known exactly, therapy is inadequate to obtain restitutium ad integrum or stop the disease and prevention—limited to the identification and reduction of risk factors which are not the cause but only a prediction of the disease—has not greatly reduced morbidity.

## Facts Supporting Adrenergic Stress in CHD

In our postmodern society "stress" is the leitmotiv of our concern. It is the major risk factor for CHD in terms of psychological tension or nervous strain. We know that the autonomic nervous system regulates body functions by a balance between sympathetic and vagal innervation. The old distinction between "Type A" and "Type B" personalities with a physiological prevalence of one of the two systems was and still is a valid discrimination, even if it is difficult to prove objectively any distinction between people who are more prone to be sympathotonic or vagotonic (Friedman et al, 1973). Nevertheless, there is in increasing interest in the role of the adrenergic system in cardiology and in CHD in particular (Leenen, 1999). In the latter the possibility and role of a sympathetic storm is apparenlty supported by the following:

1. the Cardiac Arrhythmia Suppression Trial was prematurely terminated because of a high mortality in patients not receiving beta-adrenergic blocking agents (Peters et al, 1994);
2. reduction of sudden death amongst patients with ischemic heart disease treated by lipophilic beta-blockers, which easily pass the blood-brain barrier increasing cardiac vagal tone and electrical stability of the heart (Wikstrand et al, 1992);
3. clinical demonstration of sympathetic overactivity, i.e., significant relation between the number of ischemic episodes or the overall duration of silent ischemia and norepinephrine spillover both at rest and after cold test in active unstable angina in inactive unstable angina, stable effort angina and controls (Schwartz et al, 1984, 1992; Neri Serneri et al, 1993; McCance et al, 1993); and impairment of baroreflex modulation of vagal and sympathetic outflow by abrupt angioplastic occlusion (Airaksinen et al, 1998);
4. a higher frequency of sudden death in patients with a heart rate ≥65 beats per minute (indicating low parasympathetic activity) vs. ≤65 beats without relation with other risk factors (Algra et al 1993);
5. the damaging effect of cigarette smoking because of sympathetic outflow (Narkievicz et al, 1998; Domburg et al, 2000);
6. the stimulation of the adrenergic system by cocaine (Vongpatanasin et al, 1999), in congestive heart failure, coronary heart disease (Du et al, 1999) and brain injury (White et al, 1995, Baroldi et al, 1997); and
7. experimental reduction of ventricular fibrillation threshold by beta-blockers (Baroldi et al, 1977; Anderson et al, 1983) likely related to calcium influx (Clusin et al, 1982). Beta-blocker agents infused into a conscious dog with coronary narrowing lessened myocardial dysfunction and abnormal flow patterns (Tomoike et al, 1978).

The adrenergic system, fully developed in mammals, is an adaptative system which allows rapid adjustment in heart rate, contractility, airways, peripheral resistance, venous tone, glycogenolysis, platelet adhesiveness, coagulability, etc. It is an alarm system which permits the animal to react quickly and effectively to acute situations either by aggression, escape or other emergency reactions (Cannon, 1942; Raab, 1970; Kubler, 1992, 1994). In coronary heart disease the sympathetic system may go out of control and aggravate myocardial damage by inducing progressive cell injury and ventricular tachyarrhythmias. Myocardial ischemia is a unique situation of increased sympathetic stimulation with increasing neurotransmitter release, diminished activation and regulation of receptors and temporary activation of effector enzymes leading to extension of infarct size and induction of threatening ventricular arrhythmias (Kubler, 1992, 1994). We emphasize that sympathetic overdrive is not due to ischemia and rather than "extension of infarct size", we should speak of an "adrenergic wavefront", CBN being the morphologic hallmark of an out-of-control sympathetic response. The question is whether in any case of sudden death in which this morphologic hallmark is obvious ("mural" coronary artery, Morales et al, 1980; isolated congenital coronary artery anomalies, Taylor et al, 1992; right ventricular arrhythmogenic dysplasia, Thiene et al, 1988), we may speak of a similar sympathetic storm, no matter what its cause.

In this context, sudden death cases without "critical" coronary atherosclerosis must be mentioned. In our experience, amongst 28 cases (all SD 1st episode) with a normal coronary lumen or a lumen reduction less than 50%, the frequency of CBN was 78%. This suggests that a sympathetic storm could occur in the absence of severe coronary atherosclerosis. In turn, it emphasizes a need to study the whole sudden death population without selecting only those cases with severe coronary damage and second, the inappropriateness, in forensic medicine practice, of linking sudden coronary death to severe coronary atherosclerotic obstructions alone. Furthermore, sudden death associated with malignant arrhythmias in any coronary condition, e.g., congestive heart failure in chronic coronary ischemia, (Lesch et al, 1984); in animals with sequential intracoronary embolization, (Sabbah et al, 1992) and in human non coronary conditions e.g., familial hypertrophic cardiomyopathy, (Anderson et al, 1983; Hecht et al 1993); myxomatous mitral valve, (Pasternac et al, 1982; Chesler et al, 1983); systemic amyloidosis, (Falk et al, 1984); congestive heart failure, ( Packer, 1985; Parmley, 1987); cardiomyopathies (Goodwin et al, 1976,1978; Maron et al, 1978; Marcus et al, 1982; Oparil et al, 1985; Brandenburg, 1985; Hecht et al, 1993); radiation-induced coronary obstruction (Angelini et al, 1985) or systemic sclerosis (Bulkley et al, 1978) merits further investigation to establish the presence of histologic signs of catecholamine cardiotoxicity. The latter were present in sudden and unexpected death in a case of acquired immunodeficiency syndrome with lymphocytic myocarditis (Baroldi et al 1993) and of seropositive subjects for Chagas' disease, without a subjective or clinical history of cardiac disease but with extensive myocarditis (Baroldi et al, 1997). In this context it is interesting to note that in both myocardium and diaphragm, muscle contraction band necrosis has been observed in sudden infant death (Silver et al 1992,1996).

Two aspects seem pertinent in this line of reasoning. One is the occurrence of myocardial denervation in cardiac diseases and the other is the interrelation between heart and brain.

### *Myocardial Denervation*

The pathology of the intramyocardial nerves and their endings in man is practically unknown. Few studies deal with innervation in animals (Paessens et al, 1980) or epicardial and endocardial innervation in humans (Marron et al, 1995). The intrinsic cardiac nerves form a very complex interactive regulatory system which includes afferent neurones, local interconnected neurones and both sympathetic and parasympathetic efferent postganglionic neurones (Armour, 1999). This system is affected in many cardiopathies, particularly those associated with sudden death; as, for instance, in the autonomic neuropathy in diabetic patients with increased sympathetic tone (Jacoby et al, 1992; Di Carli et al, 1999).

At present, the major sources of information are scintigraphic studies by iodine-123-metaiodobenzylguanidine (MIBG) and other in vivo methods which show a reduction or denervation of adrenergic nerves endings in experimental and human infarction (Barber et al, 1983; Stanton et al, 1989; Dae et al, 1991; Spinnler et al, 1993; Kramer et al, 1997); in adjacent noninfarcted regions (Matsunary et al, 2000); in both ischemic heart disease (Calking et al, 1993) and its absence (Mitrani et al, 1993); in relation to malignant arrhythmia (Schaeffers et al, 1998); in vasospastic angina (Kaski et al, 1986; Sakata et al, 1997; Takano et al, 1997); in arrhythmogenic cardiomyopathy (Wichter et al, 1994; 2000); in dilated cardiomyopathy of unknown origin (Schofer et al, 1988; Ungerer et al, 1998); in congestive heart failure (Bean et al, 1994; Kaye et al, 1994; Nakata et al, 1998); in Chagas' disease (Emdin et al, 1992); and in inherited ventricular arrhythmia (Dae et al, 1997). The severity of denervation and the relationship between neural and humoral stimulation are related to the magnitude and dispersion of local repolarization (Yoshioka et al, 2000). The recovery of function of a stunned myocardium by catecholamine infusion but not by stellate ganglia stimulation after 25 minutes of coronary occlusion suggests a destruction of nerves within the stunned myocardium (Ciuffo et al, 1985).

Denervation may be due to different factors, e.g., asynergy with tensile mechanical forces, interstitial inflammation, medial neuritis and so forth. No matter what its cause, denervation, particularly when focal, can trigger myocardial supersensitivity to catecholamines resulting in fatal malignant arrhythmia (Donald, 1974; Bevilacqua et al, 1986; Inoue et al, 1987; Rundqvist et al, 1997) as in diabetes mellitus (Stevensen et al, 1998), in congestive heart failure (Cao et al, 2000) and Chagas' disease which in turn are associated with catecholamine contraction band lesions (Baroldi et al, 2001). At present, the likelihood and timing of zonal reinnervation is not known. In a "model" of global denervation, i.e., the orthotopical transplanted heart, the process of reinnervation is still unclear (Wilson et al, 1993; Halpert et al, 1996), being late and dyshomogeneous (De Marco et al, 1995), and requiring 15 years to be completed (Bengel et al, 1999). Reinnervation of a focal denervation, if it occurs, may need a shorter time. In this context, the nature of dysfunction and recovery of a viable myocardium (Udelson et al, 1994) is still a matter of speculation and denervation must be considered. Of some interest would be to establish by MIBG if plaques fractured by angioplasty produce denervation of dependent myocardium by medial nerves destruction. Finally, a dysfunction, equivalent to focal denervation, but without nerve destruction, might also occur from an impaired turnover of catecholamines in the interstitium, a possibility to be considered since there is a decreased noradrenaline uptake in disarrayed but not in normal myocardium in patients with hypertrophic cardiomyopathy (Li et al, 2000) or at the border of a cardiac aneurysm (Bevilacqua et al, 1986).

### *Heart and Brain Relationship*

The heart/brain relationship (Lown, 1977, 1979; Skinner, 1985) is another matter for investigation. In acute brain lesions focal CBN necrosis is described (Connor, 1969) and prevented experimentally by a beta-blocker (Hunt et al, 1972). In experimental subarachnoid hemorrhage, arrhythmias develop with the sudden increase in intracranial pressure. Animals which had with both vagi and sympathetic nerves severed but had an intact spinal cord, developed delayed arrhythmias that could not be correlated with intracranial pressure. These findings suggests that arrhythmias could be produced either by direct autonomic discharges to the heart or by the proven increase of circulating catecholamines (Estanol et al, 1977). In animals with coronary occlusion psychological stress, e.g., electrical shock in a dog with or without an acute infarct, can alter the ventricular threshold of cardiac vulnerability and provoke major arrhythmias (Corbalan et al 1974). In patients with silent CHD, mental stress induced myocardial ischemia (Kral et al, 1997).

A more precise histological documentation is needed for all cases of different behavioral types (Kahn et al, 1987) who die suddenly following emotional distress. "Myofibrillar degeneration" has been observed in victims of assaults who had no internal injuries (Cebelin et al, 1980), and particularly in coronary heart disease patients in whom helplessness or hopelessness were the basic feelings (Engel, 1971). In turn, this raises the still undetermined mechanism of "sudden death" in aboriginal people ("voodoo death; Cannon, 1957). Intense emotion is a common denominator of these psychological states (Lynch et al, 1977; Eliot et al, 1985) in which a loss of balance between sympathetic and parasympathetic modulation may produce tachycardia and ventricular fibrillation or bradycardia or asystole. Psychological factors have a marked impact in cardiovascular disease (Rozanski et al, 1999; Freasur-Smith et al, 2000; Angerer et al, 2000), as shown by sudden death after earthquake (Leor, 1996). The question is to document whether sudden death occurs in the presence of normal or atherosclerotic coronary arteries.

We note that "passive" emotion or fear, per se, seem insufficient to elicit sudden death. For instance, no epidemics of sudden death were reported during War World II, despite every type of numerous stressful conditions in very large groups of people of different races everywhere in the world. On the other hand, a classic pattern of clinical transmural anteroseptal myocardial infarction without infarct necrosis has been described in two patients, one with a cerebral infarct and the other with a primary brain tumor. At autopsy, only "focal myocytolysis" and severe coronary atherosclerosis were observed (Duren et al, 1976). These cases support a brain/heart relationship in man in the presence of coronary atherosclerosis and emphasize the lack of specificity of the ECG in discriminating different forms of myocardial necrosis. "Ischemic" ST depression is the ECG sign of acute CBN following experimental infusion of noradrenaline or isoproterenol in the absence of flow reduction without ventricular fibrillation despite increasing myonecrosis with increasing dosage (Todd et al, 1985). These ECG changes also had a high incidence in patients resuscitated from out-of-hospital cardiac arrest at an exercise test (Sharma et al, 1987).

## Introduction to the Natural History of Catecholamine Myotoxicitiy

For reasons given above, myocardial contraction band necrosis, no matter which of its many synonyms one cares to use, is the pathognomonic lesion of myocardial catecholamine damage linked with peroxidation. No systematic, quantitative studies of CBN in human and experimental pathology were done until we quantified this lesion in terms of number of foci and necrotic myocells x 100 $mm^2$, in hearts of diseased and normal people (Table 47; Baroldi e al, 2001), thereby obtaining the following data:

1. CBN was absent in normal people who die following carbon monoxide intoxication, electrocution and head trauma while it was present in 42% of such subjects of the latter group if they survived longer than one hour. All normal subjects died out-of-hospital, did not receive any therapy, had negative autopsy findings, their coronary arteries being normal or with insignificant atherosclerotic plaques.
2. A statistically significant increase in both the frequency and extent of the lesion was recorded in acute infarct cases, transplanted hearts, sudden/unexpected coronary and Chagas's death, and those dying of intracranial hemorrhages compared with other groups.
3. There was a higher frequency and extent of CBN in people who survived longer.
4. CBN had no relation to resuscitation therapy. The same extent of "early" lesion was observed with or without the latter.
5. In groups with a greater extent of CBN all morphologic stages of the lesion from early hypercontraction/rhexis to healing and healed phases were observed.
6. CBN is not an artifact or a postmortem change and can not be confused with reflow necrosis.

Our conclusion was that in the agonal phase, an adrenergic stimulus to increase contractility seems to be present. In this condition, the earliest CBN change and its minor extent indicate a prolonged terminal event of great significance to pathologists. In other specific diseases where CBN is found, an adrenergic stress is part of their natural history. The threshold for a diagnosis of adrenergic overactivity seems to be an extent of CBN beyond the range seen in more involved groups (Table 47) particularly when associated with the coexistence of different morphological stages of the lesion. Overall, the amount of tissue loss is insignificant in term of reducing myocardial contractility. Contraction band necrosis however, is an important hallmark of adrenergic stress with arrhythmogenic supersensitivity. It is a hallmark to be checked and quantified in many, if not all, cardiac and noncardiac diseases.

Schlesinger and Reiner (1955) described a loss of myocardial cell cytoplasm, with the cells reduced to empty sarcolemmal tubes filled by macrophages loaded with yellow-brown pigments as "myocytolysis". The authors indicated that myocytolysis healed by collagen fiber deposition without the formation of granulation tissue. They found the lesion in 64% of acute-recent infarct cases, in 16% of hearts with myocardial fibrosis and in 2% of cases in the absence of myocardial damage. No explanation for this lesion was provided. Subsequently, at the edge of acute infarcts, Saram (1957) described a "fibrillolyse", i.e., a dissolution of myofibrils in edematous myocardial cells, without any cellular reaction or repair process. In our first contribution to the subject, (1975) Saram's finding was confirmed with the lesion being present in the subendocardial and perivascular myocardial layer, usually preserved in a myocardial infarct. Then, it became clear to us that "myocytolysis" as described by Schlesinger and Reiner, was a late, repair stage of contraction band necrosis found not only surrounding an infarct but also in nonischemic myocardium of the infarcted heart and in noncoronary conditions (Baroldi et al, 2001). To maintain the use of "myocytolysis" so that it indicates fundamental damage to myofibrils, we made a distinction between coagulative myocytolysis—named because segments of hypercontracted/coagulated sarcomeres formed bands in the myocells—and colliquative myocytolysis in which there was myofibril colliquation in edematous myocardial cells. We also investigated the frequency and extent of colliquative myocytolysis in the same material studied for CBN (Table 48). A significantly higher frequency and extent of this lesion was observed in hearts with congestive failure versus all other groups ($p<0.001$). We note and decry the common practice of interpreting both colliquative and coagulative myocytolysis as ischemic lesions. We reemphasize that coagulative myocytolysis is a catecholamine-induced lesion. The cause of colliquative myocytolysis is less clear. However, its relation to heart failure and its absence in any stage of myocardial infarct necrosis and in experimental reperfusion necrosis speak against ischemia as a possible cause (see above).

## Adrenergic Stress and Related Morphologic Changes

We must consider two aspects of adrenergic stress in cardiac disease, one that produces structural changes (including both nerve and myocardial pathomorphology) and another involving ionic and molecular fluxes invisible, morphologically. At present, we have only indirect morphological evidence that adrenergic stress occurs in different situations, associated or not with coronary atherosclerosis. Nevertheless it is clear that contraction band necrosis is observed in cases of sudden death if they are looked for assiduously. Recently, examining three consecutive young subjects, all of whom died suddenly and unexpectedly and had no clinical history of any disease and with normal coronary arteries, the unique findings were CBN and myofiber breakup. The latter may be associated with ventricular fibrillation as ECG demonstrated terminally in one case.

These episodic examples, however, show the need to quantitate all morphologic findings in cases of sudden death to establish both their significance and the real meaning of adrenergic stress in cardiac diseases. We emphasize (Table 47) that cardiopulmonary resuscitation per se,—including noradrenaline infusion, electrical defibrillation, etc—did not seem responsible for catecholamine damage in the cases we studied, at least when the latter had the "threshold" previously reported. Higher value and the presence of different stages of the lesion seem an acceptable postmortem indicator of adrenergic stress repeated over time. On the other hand, sympathetic overstimulation may act on several different targets when one considers the many effects of the adrenergic system on, for example, blood coagulation, platelet adhesiveness, smooth muscle cell hyperplasia, arterial wall tone, hyperlipidemia, denervated myocardium etc (Eisenberg, 1966; Levin et al, 1964; Kaplan, 1987, 1988; Yamori et al, 1987; Cruickshank et al, 1987; Ablad, 1988; Wikstrand et al, 1988; Velican et al, 1989; Ross, 1993; Leenen, 1999) and its linkage with immunologic responses in relation to cardiovascular changes (Benshop et al, 1994; Maisel, 1994). The adrenergic storm may occur in many conditions with overexpression of β-receptors associated with myocardial fibrosis and heart failure (Ligget et al, 2000), be induced by free radicals and prevented by beta-blocking (Flesh et al, 1999; Fineschi et al, 2001) and possibly be related to oxidative stress (Buffon et al, 2000). In coronary heart disease, in particular, the role of sympathetic overstimulation could explain (a) the starting point, i.e., nodular myohyperplasia and the progression of the coronary myohyperplastic atherosclerotic plaque in which process the interactive role of hemodynamic wall stresses/neurologic control of both coronary circulation and heart pump have a role; and (b) both complications and cause of death.

Other morphological changes might be included amongst adrenergic-related effects and merit further investigation. One is myocardial disarray and its linkage with catecholamine,/ endocrine disorders (Goodwin, 1982). This linkage is supported by several experiments showing a catecholamine increase following administration of nerve growth factors (Witzer et al, 1976); triac (acetic analogue of tri-iodothyronin) in pregnant rats with disarray in fetuses (Hawkey, 1981) and its prevention by beta-blocking; or subhypertensive doses of norepinephrine (Laks, 1975; Blautass, 1975); the cardiac directed overexpression of human beta 1-adrenergic receptors in transgenic mice leading to myofibrillar disarray, marked cardiac hypertrophy and interstitial myocardial fibrosis plus cardiac dysfunction and sudden death in older animals (Bisognano et al, 2000). In our experience "pathologic" (i.e., ≥ 20% of the histological area) disarray was observed in conditions where adrenergic stress is admitted. In Table 49 are listed all conditions we had the opportunity to investigate systematically. The frequency and extent of disarray were statistically more significant in intracranial brain hemorrhage, sudden/unexpected coronary death, transplanted hearts, sudden/unexpected death in silent Chagas' disease and cocaine abusers. What is interesting to note is how disarray was absent in transplanted hearts when survival was less than 7 days. This suggests that time is required to develop this form of architectural distortion, which in turn could be an arrhythmogenic trigger.

The other structural change is endocardial myoelastofibrosis. In contrast to the current opinion that endocardial thickening is secondary to the healing of mural thrombosis, we showed that the latter is a relatively rare event and that endocardial thickening is the end result of a process which starts as nodular smooth muscle cell hyperplasia followed by elastic fiber proliferation and fibrosis. This is a process which mirrors the earliest stages of the atherosclerotic plaque. Can this change, which occurs in different conditions (Table 50), be related to adrenergic stress?

## Adrenergic Stress and the Etiology of CHD

We have presented our viewpoint in interpreting the sequence of events observed in the natural history of acute and chronic coronary syndromes. Our morphologic data oppose the concept that ischemia is the prime cause of the latter, showing that each single step in any pattern of this disease has its own pathogenesis and that each single pathological change has to be revisited in terms of dysfunction. Rather than ischemia, we believe that the prime cause is a molecular/ion disorder of a small or large focus of myocardial cells triggered by the autonomic nervous control, particulary of the adrenergic system. In this respect a distinction must be made between blood-borne catecholamines and that released within the myocardium. Already we mentioned that the systemic infusion of catecholamines which produces diffuse CBN in the whole heart does not determine arrhythmia/ventricular fibrillation, rather the latter occurs following infusion in one coronary artery. This is in line with the regional changes characterizing the acute coronary syndromes. The unique morphologic finding which may have some meaning in relation to adrenergic nerves is coronary medial neuritis, topographically located at the plaque level only and related to its proteoglycan accumulation phase. The statistical significance of the frequency and extent of medial neuritis in CHD patients versus normal controls confirms a cause-effect association between coronary atherosclerosis and CHD according to the following working hypotheses, to explain (a) local myocardial hypokinesis prior to an infarct; (b) ventricular fibrillation as a mechanism of sudden cardiac arrest; and (c) myocardial failure in CHF. With the need to explore the cause/effect relationship when similar conditions occur in the absence of atherosclerotic coronary disease.

Degeneration of inflamed adventitial-medial nerves may result in an area of myocardial denervation with it becoming hypokinetic and at high risk of stretching any time increased intraventricular pressure occurs with secondary ischemia due to extravascular compression resulting in infarction. AMI cases had a significant highest frequency and extent of medial neuritis. This hypothesis pertains to atherosclerotic coronary disease and cannot explain the infarct cases without coronary atherosclerosis. Little is known about how local myocardial denervation occurs and its influence on contractility. For instance, we hypothesized that sudden and unexpected death in Chagas patients without clinical manifestation was due to intramyocardial nerve destruction secondary to extensive and diffuse foci of lymphocytic inflammation which was associated with contraction band necrosis secondary to sensitization of denervated myocells to catecholamines (Baroldi et al, 1997). On the other hand the nervous structure within asynergic myocardium, irreversibly stretched or hypercontracted, can be destroyed by the mechanical force of the contracting myocardium acting on the asynergic one. The recovery of a stunned myocardium following catecholamine infusion, but not by stimulation of stellate ganglia (Ciuffo et al, 1985), supports this viewpoint.

Irritation of these nerves may cause (a) an excessive release of noradrenaline within the myocardial interstitium (Lameris et al, 2000) as an arrhythmogenic factor causing sudden death in the absence of a myocardial infarct; or (b) a coronary spasm capable of abolishing any collateral flow with temporary ischemia followed by regional dyskinesis and adrenergic overstimulation by mechano-receptors again resulting in sudden death. We favor the first hypothesis since most SD cases are not infarct related.

On this subject, the hypothesis that a spasm of resistive intramural arterial vessels caused by endothelial vasoactive substances is difficult to prove and to distinguish this spasm from extravascular compression. Furthermore, one should explain what triggers a specific and often only a small area of the endothelial vasomotor activity, in contrast to the enormous endothelial surface which include veins and lymphatics.

The fact of adrenergic overactivity in congestive heart failure likely related to angiotensin (Heneger et al, 1998), is generally admitted and beta-blocking agents have a positive therapeutic effect (Cohn et al, 2000; Mortara et al, 2000). Again, CHF represents a primary myocardial cell disorder occurring in a variety of conditions. The contraction state of apparently normal myocells, as shown ultrastructurally, suggests that the diminished compliance of the contraction cycle is due to a reduced capability to relax (Piper et al, 2000). A disorder likely related to a defect of the $Ca^{++}$ pump to remove the ion from the troponin-tropomyosin complex and possibly linked with an increased catecholamine turnover. The effect should be an abortive hypertrophy with an altered sarcomerogenesis and longitudinal formation of new sarcomeres without increased cellular volume, thus explaining the weight/size paradox. A working hypothesis is supported by: the increased length of the myocardial cells formed by sarcomeres of normal length and a greater cell length/width ratio seen in cultured myocardial cells sampled from failing ischemic hearts excised at transplantation (Gerds et al, 1992); by $Ca^{++}$ intracellular content (Bakker et al, 1995); the deviation of normal myofilament $Ca^{++}$ sensitivity important in the mutation of sarcomeric proteins (Michele et al, 1999); and systolic and end-diastolic velocity's strong dependence on both number of myocytes and density of myocardial beta-adrenergic receptors (Shan et al, 2000). With two observations (a) the abortive hypertrophy is a reversible phenomenon as shown by the benefits of beta-blocking therapy and by the optimal functional recovery of a severe CHF in patients with malformed heart after surgical repair; and (b) in congestive heart failure we are dealing with an increase of blood-borne catecholamines. In fact, as in their experimentally intravenous infusion or in pheochromocytoma, in hearts with congestive failure CBN foci of any age are found in the whole myocardium.

## Concluding Remarks

In a past editorial Baroldi (1978) recalled that "for millennia the earth remained the center of the universe; an unquestionable, objective fact documented already from the first man by looking every morning from his cavern, at the rise of the sun and its course from East to West. A few, by looking at the stars, questioned this undeniable fact and, despite the Inquisition, were right" (Baroldi, 1978). However the truth was recognized after almost 800 years from Aristarco (270 BC) to Copernico and Galileo (1550 AC). Today, dogma provides another supposedly undeniable fact viz—it "is hardly credible that there should be continuing debate about what is ostensibly so simple a morphologic problem: the relationship of coronary thrombosis to acute myocardial infarction" (Davies et al, 1976). This has in turn promoted and promotes sophisticated, highly technological, preventive and therapeutic procedures, ranging from manipulation of gene expression that will stop plaque rupture, to new vessel growth to vascularize myocardial scar and hibernating myocardium to implantations of new myocardial cells to resolve insufficient contractility, all under the aegis of the "thrombocentric" CHD universe. These are fascinating perspectives for the new millenium amongst which the old Virchowian concept on the inflammatory nature of atherosclerosis ("chronic endoarteritis deformans; 1856) is revitalized by (a) a relation between the severity of the latter and antibodies against *Clamydia pneumoniae* (Ericson et al, 2000); or (b) by C-reactive and amyloid A proteins as generic indicators of inflammation related to higher mortality (Liuzzo et al, 1994, 1998; Morrow et al, 1999); or (c) monocytic/granulocytic activation in CHD (Neri Serneri et al, 1992; Mazzone et al, 1998). We were able to see medial neuritis only at the proteoglycan stage of the atherosclerotic plaque as an unique (autoimmune?) process in the natural history of coronary heart disease. The impression is that CHD needs to be reinterpreted, or, unfortunatelly, it will remain an important if unknown "universe" jeopardizing our evolution.

One can paraphrase Hurst's sentence (1967) "Some words prevent learning" into "some morphologic findings prevent learning". Only when the functional meaning of morphological findings in this disease are interpreted correctly and their meaning resolved can we begin to seek truth. What may seem so simple does not reflect fact. Sudden death/coronary heart disease appears to be an integrated set of phenomena of which plaque fissuring is only one (misinterpreted) morphologic finding. We present above much morpho-functional and clinical data that make us doubt the value of the currently accepted simplistic "unifying theory" and the cause/effect relationship between an occlusive coronary thrombus and all forms of myocardial necrosis. Rather, we believe that morpho-functional and clinical data support a secondary role for plaque rupture, thrombosis and embolism and that metabolic mechanisms rather than ischemia explain complications and deaths in acute and chronic ischemic syndromes (Fig. 25). Amongst morphologic variables we have listed all those which may allow a reinterpretation of the natural history of our epidemic.

Colloquially in defining the epidemic of coronary heart disease or acute coronary syndromes, the term "coronary" has become synonymous with ischemia and is used to define an atherosclerotic stenosing-occlusive lesion that is responsible for all clinical patterns. The present review suggests a different meaning. In particular, silent or not silent ischemia is not a major determinant of an acute coronary event. Death is frequently triggered by a ventricular premature depolarization with a preceding short-long heart cycle that likely produces dispersion of refractoriness in an arrhythmic substrate (Gomes et al, 1989); and complications and death seem more related to myocardial catecholamine "toxicity" than ischemia while its onset may or may not be caused by the latter. In reality and in contrast with other opinions (Ambrose et al, 1992), each acute syndrome presents a specific morphology and clinical pattern which makes each an entity, irrespective of whether one may transform into another. We presented facts which justify why we prefer coronary heart disease rather than ischemic heart disease (Fig. 26).

Our data and concepts are an invitation to reconsider the linkage between atherosclerosis and the current epidemic of coronary heart disease; avoiding premature and misleading theories and proposing only well grounded working hypotheses. Why and how, in particular socio-economical conditions, atherosclerosis becomes the morpho-pathologic background of disease in some organs of some individuals is still a matter for research. Any idea or concept based on solid facts merits further investigations, to substitute preconcepts with truth.

As pathologists we make a plea for a high autopsy rate to allow continuing detailed studies of acute and chronic coronary syndromes. Pathology, before the astonishing and welcome development of clinical technology in the past 50 years, was the basic reference to resolve etiopathogenesis of diseases and an autopsy the main tool for a correct diagnosis. A crescendo of clinical imaging techniques, e.g., angiographic, scintigraphic, etc, has induced a continuing decrease in autopsy demand. On the other hand, pathologists are often unable to reply to clinical questions because they lack specialized knowledge to interpret clinical imagings. This suggests modifications are needed in training programs (Baroldi, 1993). In describing the natural history of any morphologic aspect within the heart—at least those we are able to see—we have shown how deep is the gap between clinical approaches and the complex reality of the structural, evolutive changes beyond vision in vivo (Baroldi, 2001). Without that information and a multidisciplinary collaboration applied to the problem of coronary heart disease, each of us, studying the condition, is like blind men examining parts of an elephant. The study of man is man. Each investigator provides snippets of information that must be synthesized to achieve full understanding; with the compulsory need for permanent contact between competent pathologists and clinicians.

This book has been written with three concepts in mind. The first is Donald's (1974) that "the value of any investigation may lie more in the questions it raises than in those it answers". The second was to list only solid facts which may promote new ideas and research protocols that lead to a better understanding and third is that any present pathogenetic theory of coronary heart disease is just that, a hypothesis. Progress to ultimate knowledge and disease prevention requires an accretion of facts and an intellectual challenge in which both agonists and antagonists are needed. Having in mind that coronary heart disease is the number one problem in health care.

# ILLUSTRATIONS

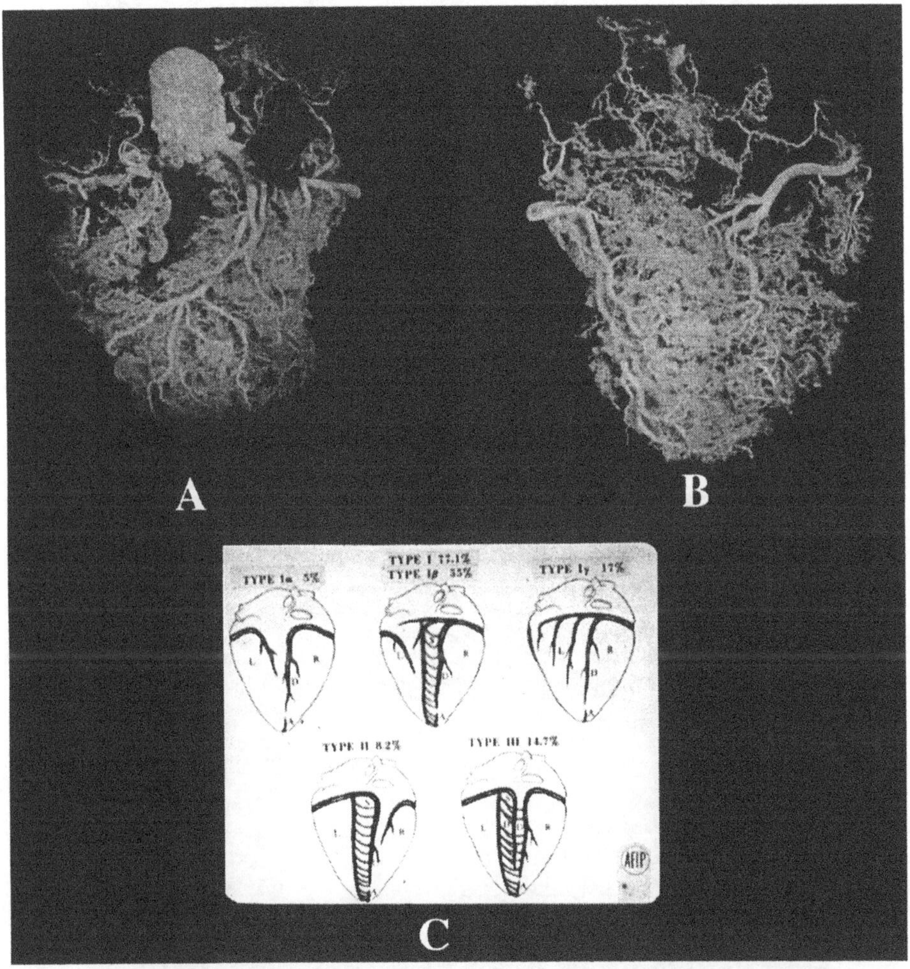


Figure 1. Tridimensional plastic cast of coronary arteries injected via the aorta. (A) anterior view with the cast of ascending aorta and (B) posterior view (C) Types of anatomical distribution mainly based on the different course in the atrio-ventricular groove of the posterior right coronary artery. Reprinted with permission from Baroldi G. In: Silver MD, ed. Cardiovascular Pathology 2nd Ed. New York: Churchill Livingstone Inc., 1991:487.

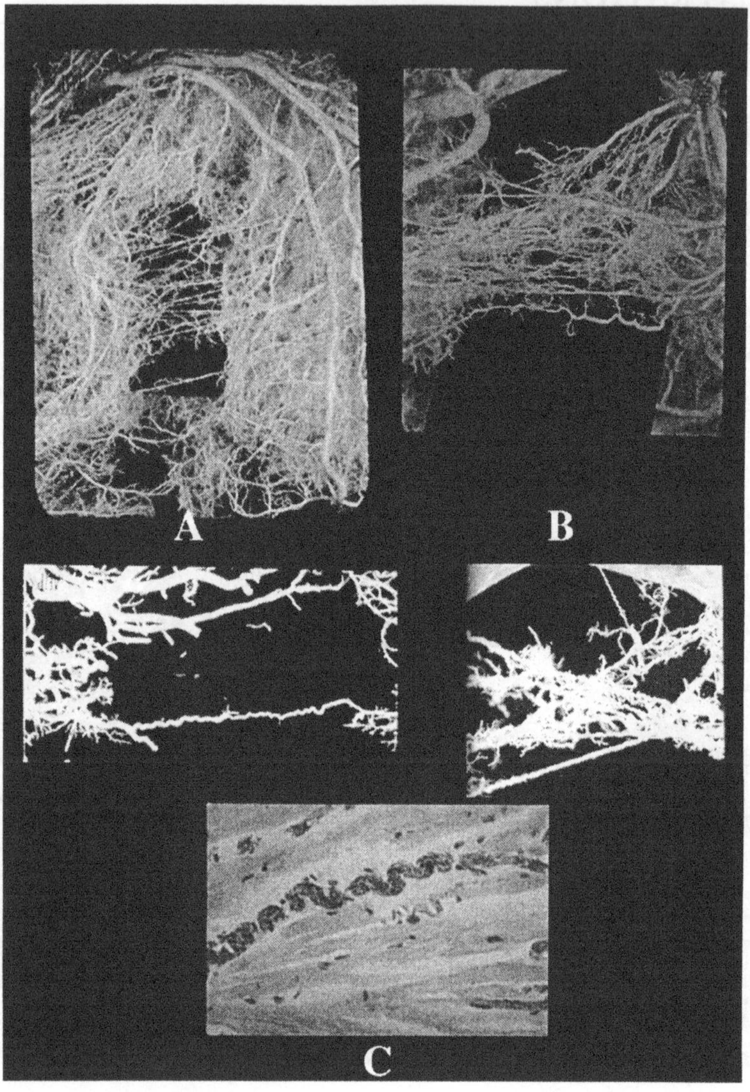


Figure 2. Coronary artery collaterals in normal human heart (A) Intercoronary and (B) homocoronary collaterals showing (C) their characteristic "corkscrew" structure related to collateral vessel disposition parallel to the line of myocardial contraction. Normal collaterals are innumerable and located at any level of the intramural arterial vessel. (D) They have a capillary-like structure (H & E x 180), are part of the terminal bed and partecipate in its delivery function.

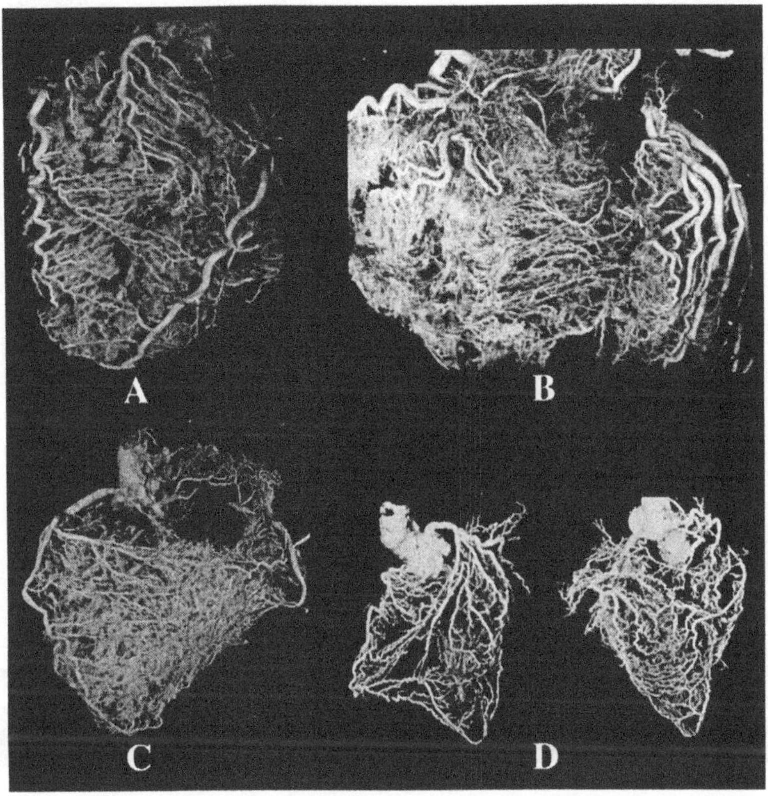


Figure 3. Increased in size of collaterals, in chronic hypoxic states and cardiac hypertrophy with normal coronary arteries and in the presence of severe obstruction of extramural arterial vessels. (A) Right ventricular hypertrophy ("cor pulmonale"). A good example of maximal hypertrophy, the right ventricle reaching the size of the left one. The extramural coronary arteries increase in diameter and length in proportion to the increased cardiac mass. No new vessel formation (B) Collaterals with increased diameter and length. A similar increase in the intramural system, including collaterals exists in cases of chronic hypoxia, e.g., anemia. (C) Occlusion of the right coronary artery (anterior view) at its origin with its distal portion filled by innumerable collaterals (D). Anterior view with superior and distal occlusion of the anterior descending branch (left) and occlusion of the right coronary artery (right) in a 62 year old man dead from brain hemorrhage without any history of coronary heart disease and histologic findings of old or acute myocardial infarct. Note "satellite" collaterals topographically related to the vascular obstruction.

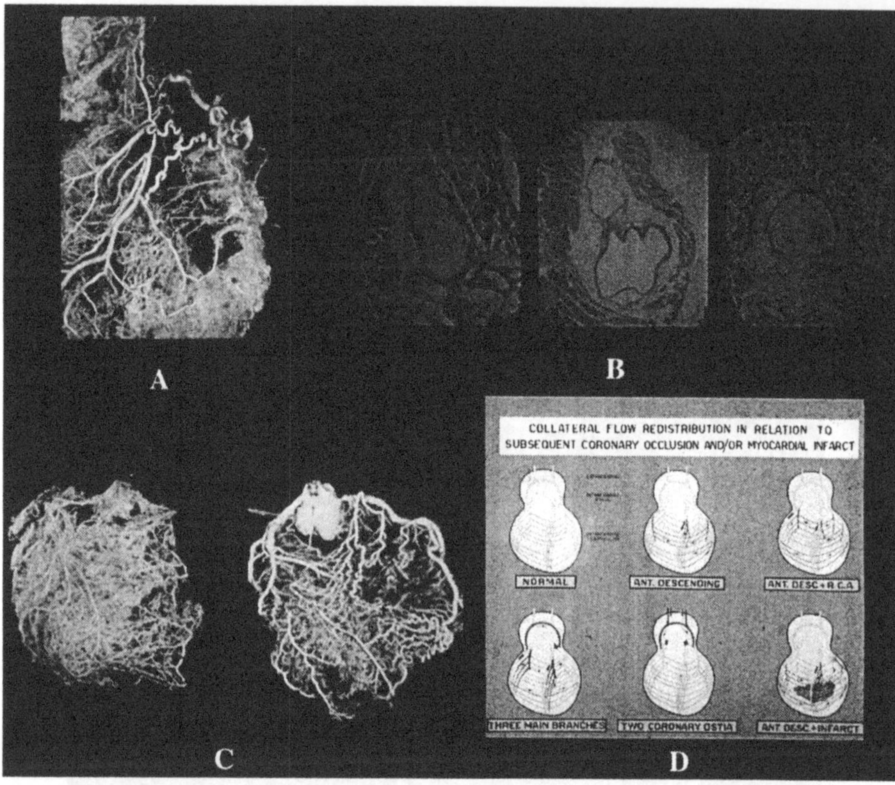


Figure 4. Compensatory flow redistribution: causes and structure of satellite collaterals (A) Occlusion of the left anterior descending branch "compensated" by a large collateral. Other normal collaterals are visible. This indicates that a pressure gradient is the stimulus for the enlargement of satellite collaterals, rather than ischemia. If the latter was the cause all collaterals of the ischemic zone should enlarge. (B) Histology of extramural collaterals sampled in the unique case with severe obstructive coronary disease in whom they were present joining subepicardial arterial vessels. Note the capillary-like wall despite a very large diameter (Mallory and H & E x 50). (C) Different aspects of collateralization. Relatively few very large collaterals vs many smaller ones in the presence of the same occlusion of the left anterior descending branch (D) Diagram showing the change in collateral flow redistribution in relation of new obstructions and following an infarct (see also Fig. 10).

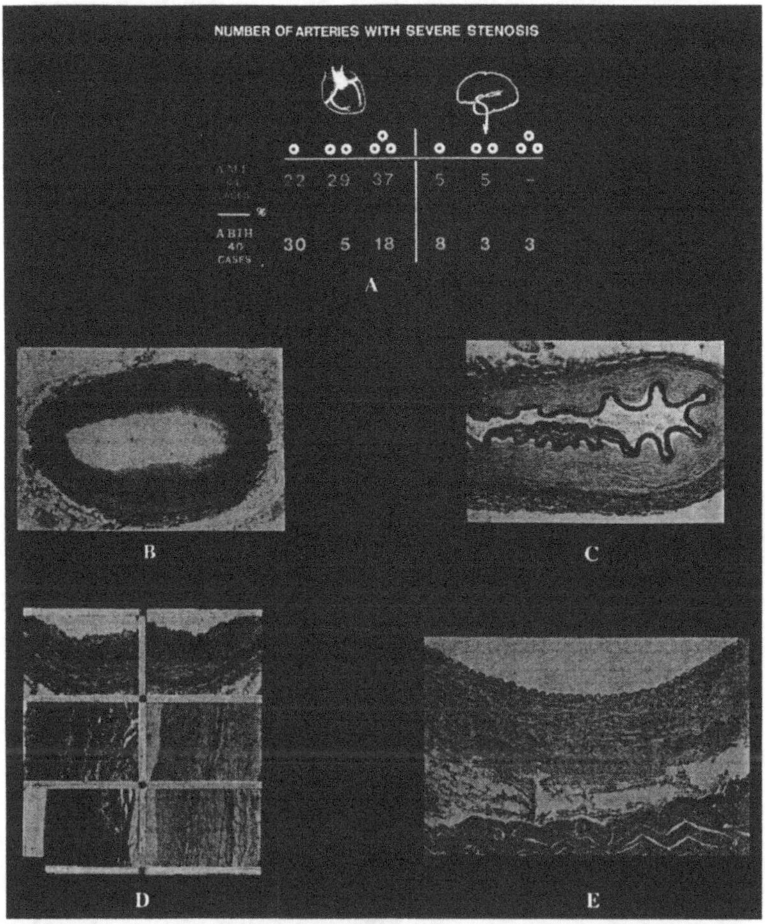


Figure 5. (A) Divergence in incidence of severe stenosis between coronary and cerebral arteries in 41 acute myocardial infarcts (AMI) in patients without brain disease and 40 acute brain infarct/hemorrhage (ABIH) without cardiac disease. Note the very high frequency of coronary and very low frequency of brain artery atherosclerotic obstructions in both groups. A similar divergence is seen for physiologic intimal thickening in (B) which shows the LAD of a 18 year old man and his (C) middle cerebral artery with minimal nodular thickening (Weigert elastic x 30 and 60). The physiological intimal thickening becomes diffuse and affects the whole surface of the extramural arteries by the end of the second decade being thicker than the media, but without altering the vessel lumen. (D) Aging of intimal thickening ranges from early myoelastic to progressively fibrous stage (Mallory and Weigert elastic x 50). (E) Absence of intimal thickening in left anterior descending branch of a dog despite a similar anatomy and diphasic coronary flow as in man. This suggests that physiologic coronary thickening in humans is due not only to diphasic flow but its neurogenic control so frequently stimulated by many psychological and physical factors.

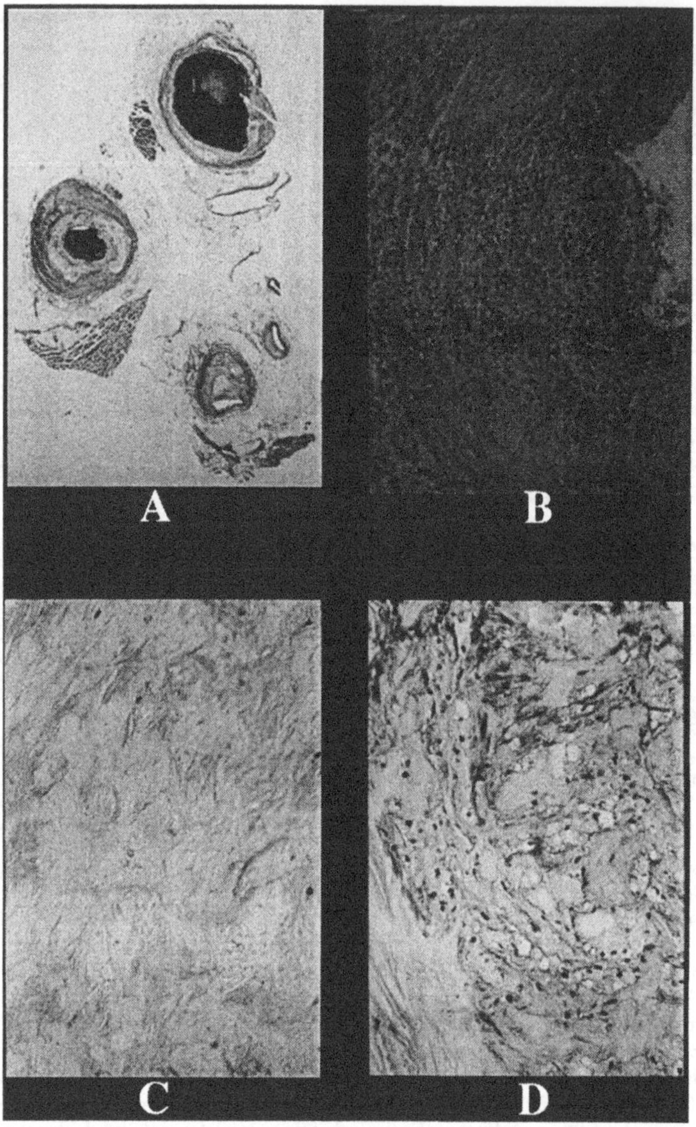


Figure 6. Coronary atherosclerosis. (A) Three different levels of an atherosclerotic plaque (top), minor nonobstructive changes with a normal lumen filled by a coagulum; (middle) severe concentric lumen reduction (≥ 70%) occluded by a thrombus with plaque showing deep proteoglycan pools + lipo-protein/cholesterol and (distal) suboccluded semilunar lumen without acute occlusion. The latter is the hemodynamically active stenosis in promoting collaterals (Movat x 15). (B) Early nodular proliferative myoelastosis (H & E x 100), with (C) proteoglycans collection deep in the fibrous intima, closed to the media (Movat x 100) and subsequent lipo-protein/cholesterol deposition (D) in part stored in macrophages ("foam cells") and/or calcium salts deposition (not shown). This sequence is typical of the "myohyperplastic" coronary atherosclerotic plaque found in both the general population and coronary heart disease patients.

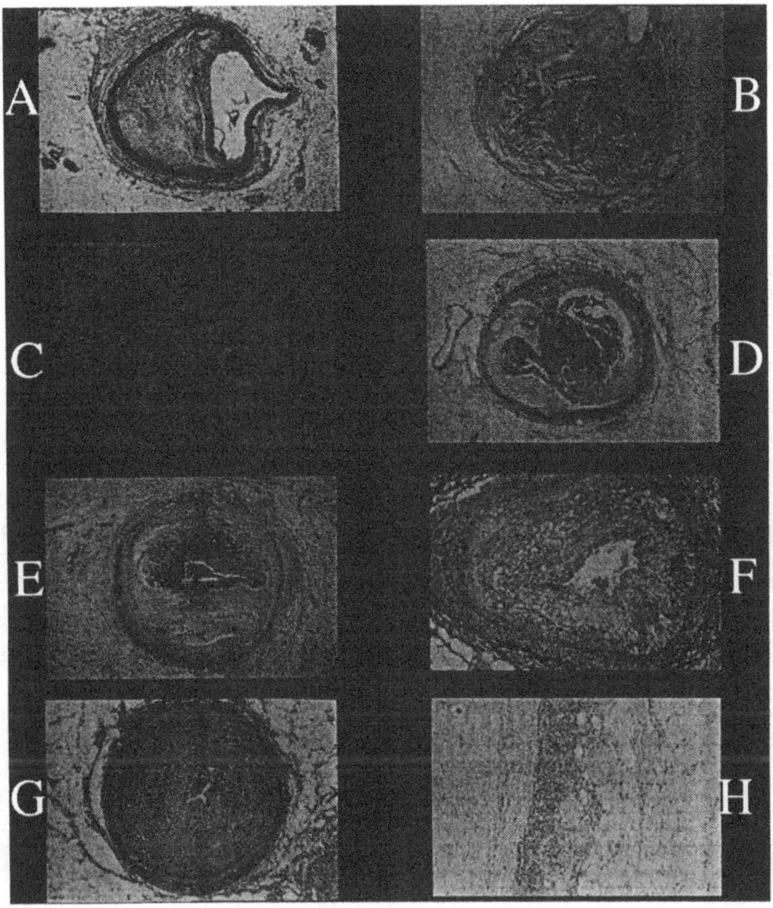


Figure 7. Different aspects of coronary atherosclerosis: (A) semilunar plaque with an almost normal lumen when distended by blood flow (Movat x 20), (B) huge "atheromatous" plaque plus hemorrhage and a very small lumen (> 95% stenosis) in central -right position (arrow) easily missed by a pathologist (Movat x 13), (C) "fibrous" plaque with >90% lumen reduction (Movat x 13), (D) pattern of an apparent plaque rupture at the site of a thrombus occluding a severely stenosed lumen (Movat x 13), while a section further downstream (E) shows a small (>90% lumen) laterally. The empty lumen (arrow) surrounded by a proteoglycan layer and a huge, hemorrhagic plaque gives the false impression of a ruptured plaque with luminal thrombosis. A pathological diagnosis may be altered by the level of section, stressing the importance of step sections through the whole plaque. Equally, to evaluate flow dynamics in a plaque, all variations in lumen reduction along its course, type of occlusion and collaterals need to be considered. (F) "Hypercholesterol" plaque mainly formed by lipoprotein/cholesterol stored in foam cells. This is the plaque found in animals fed by hypercholesterol diet and in human familial hyperlipidemia (35 year old man with type A hyperlipidemia; Weigert elastic x 20). Its progression starts with and without endothelial damage, subendothelial lipoprotein/cholesterol deposition and monocytic macrophage reaction which in our opinion is more a reparative rather than an inflammatory process. (G) Obliterative intimal thickening found in various conditions, in this case a transplanted heart of 202 days, to show its completely different morphology from both myohyperplastic and hypercholesterol plaques. Note the absence of all atherosclerotic variables and a normal internal elastic membrane; (Verhoff elastic x 50). (H) Lympho-plasma cellular inflammation around a nerve of the tunica media: medial neuritis seen only in atherosclerotic plaques parallel to the proteoglycan deposition; and is not seen in allograft coronaropathy without superimposed atherosclerosis.

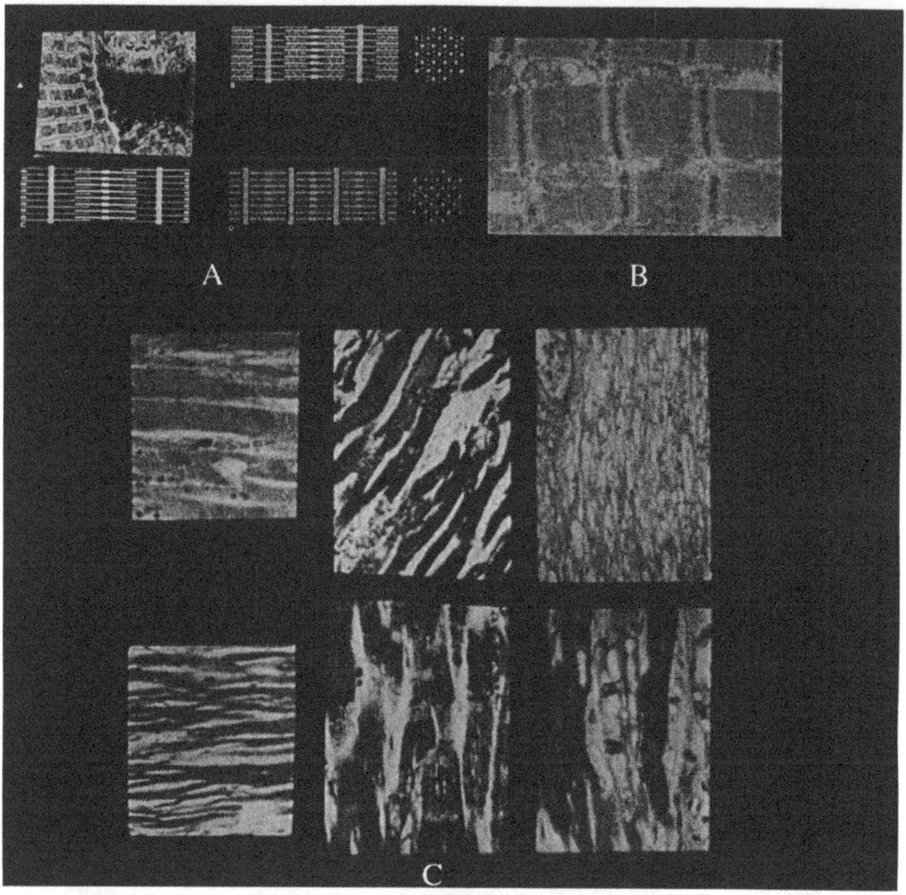


Figure 8. Rhythmic contraction-relaxation cycle. (A) Diagram showing (a) ultrastructural view of a normal myocell adjacent to a myocell with a hypercontracted pathological band (b) normal contraction state, (c) distended and (d) contracted sarcomere. (B) Normal sarcomere with all typical physiological bands (see text; EM x 40,000). The sarcomere is the functional unit of the myocardium. (C) Myocardial cells may irreversibly stop in relaxation (left) "atonic death" (H& E x 100); in irreversible hypercontraction (middle) "tetanic death" (Luna and PTHA x 100) or after progressive failure (right illustration) "failing death" (Mallory x 100, H & E x 70).

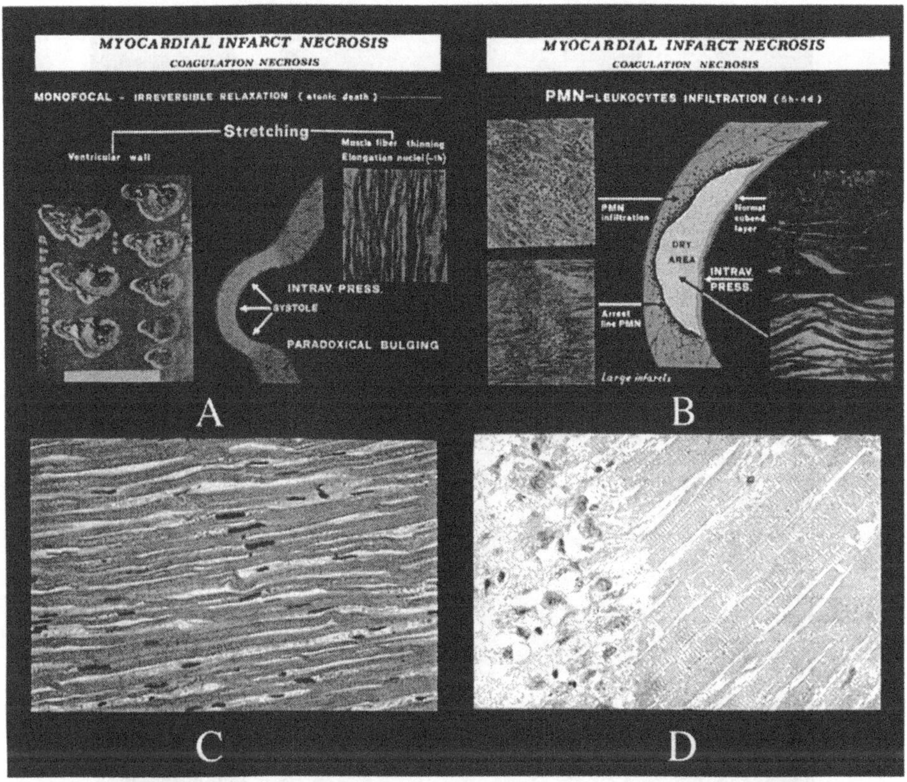


Figure 9. Infarct necrosis (atonic death). (A) Diagram showing a gross view of an infarct and the stretching of the affected myocardium with paradoxical bulging by intraventricular pressure. In turn, there is an elongation of sarcomeres and nuclei. The loss of myocardial function occurs in a few seconds of nutrient flow arrest; if the latter is not reestablished within 20 minutes the myocardium becomes necrotic and the infarct is well established within one hour. (B) Different progressive stage of this lesion with polymorphonuclear leukocytic infiltration (see text) (H & E x 250). (C) Hyperdistended myocardial cells in early AMI stage. Note the absence of vacuoles, hemorrhage, edema, pathological contraction bands (H & E x 250). (D) Removal of the necrotic tissue by macrophages without granulation tissue formation (H & E x 400). The necrotic myocells maintain their sarcomeric registered order even in the late phase of the repair.

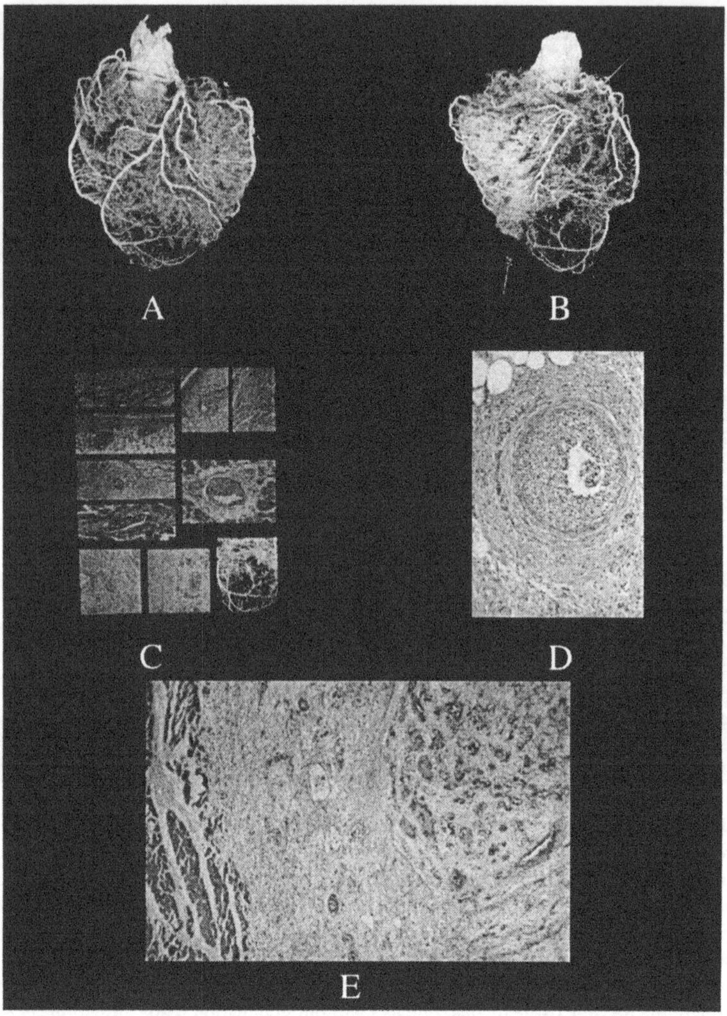


Figure 10. Infarct necrosis. Avascular area (A) anterior and (B) posterior view of a coronary plastic cast in a case of acute myocardial infarction. The latter shows a lack of injection material in intramural arterial vessels. This "avascular" area includes the distal part of the antero-posterior left ventricle and anterior interventricular septum. In this case the infarcted zone is related to a left anterior descending vessel with an almost normal lumen (only a few nonfunctional stenoses). The right coronary artery is occluded in its first posterior tract before the origin of the posterior descending branch. All arterial vessels distal to this occlusion are well filled through collaterals and no myocardial damage was demonstrated histologically in the vascular territory of the right coronary artery. Only by gross inspection, would a pathologist have diagnosed an infarct secondary to the right coronary occlusion. Development of an avascular area is an early process, already visible in the first hour. It indicates a sequestration of the affected myocardium from blood flow because of stretching of the infarcted tissue. In turn, this means it is difficult for therapeutic drugs to reach damaged tissue even in the early phase. (C) Secondary wall damage and fibrin-platelet occlusive thrombi of the intramural vessels within infarcted myocardium (H& E x 100-450). (D) Obliterative intimal hyperplasia at the margin of an infarct during the repair process (H & E x 420). (E) Lack of vascularization of a recent infarct in contrast with the new vessel growth in an associated endocardial thrombus (Movat x 125). The heart was injected postmortem. A good example to show that, when present, new vessel formation is visible.

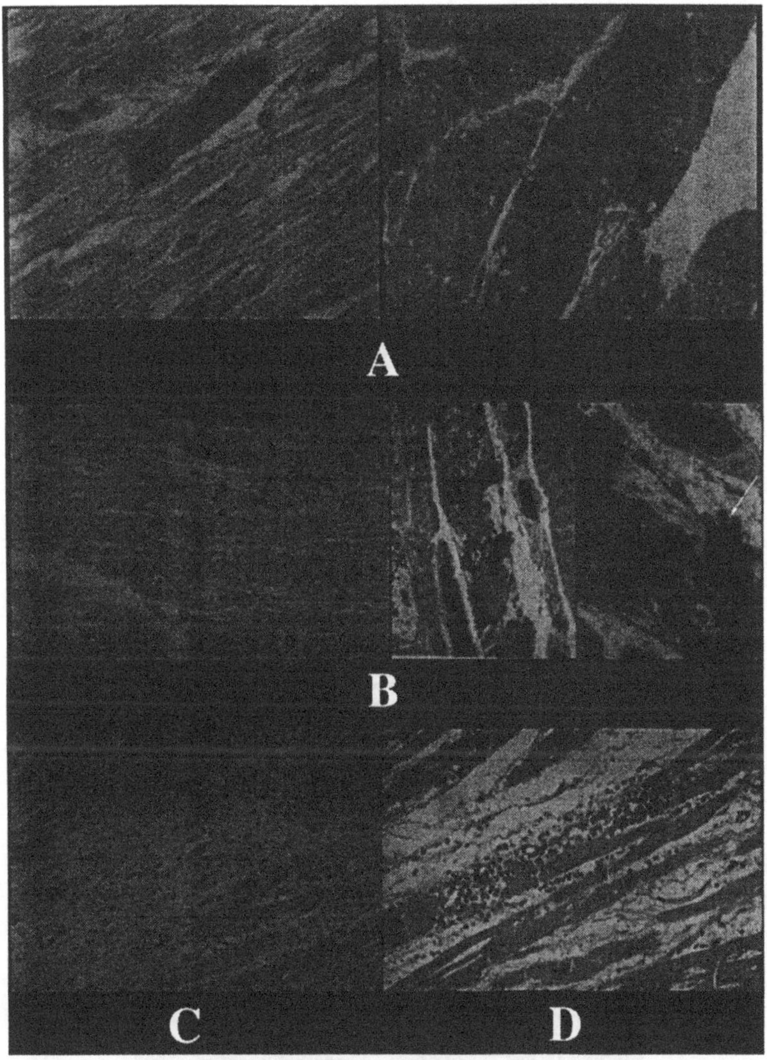


Figure 11. Coagulative myocytolysis or contraction band necrosis (CBN). Pancellular lesion. (A) Histology (H & E x 400) and ultrastructure (EM x 4,765) of a hypercontracted hypereosinophilic myocell with extremely short sarcomeres and thickened Z line. This pattern represents the first morpho-functional change after 10 minutes of experimental intravenous infusion of noradrenaline or isoproterenol. (B) A focus of CBN (H & E x 250) and ultrastructural pathologic bands with total rhexis of the myofibrillar apparatus (EM x 11,500) and coagulation of hypercontracted sarcomeres (x 23,000). (C) Healing of this lesion with the typical alveolar appearance of myocytes (empty sarcolemmal tubes) plus macrophages loaded by lipofuscin and a few monocytes (H & E x 250). When the lesion involves the whole cell we defined it as pancellular lesion. One notes that the early repair process starts with monocyte (D) and macrophage activity (H & E x 250) a pattern easily confused with that of a lymphyocytic myocarditis. Keep in mind that CBN is a frequent lesion found in many conditions, even in the absence of coronary heart disease; thus phenotype discrimination is needed for a correct diagnosis.

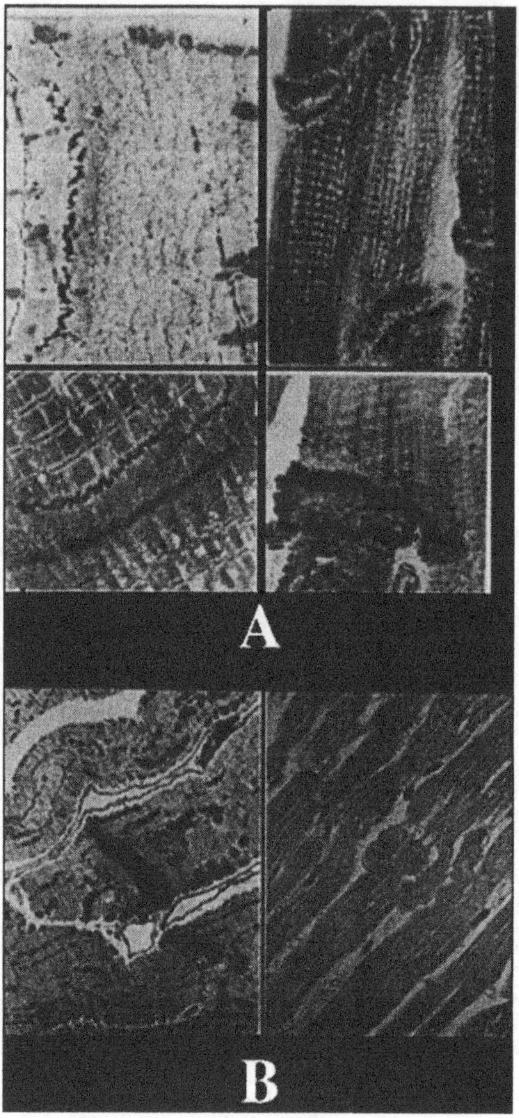


Figure 12. CBN - Paradiscal lesion. It consists of a band of about 15 hypercontracted sarcomeres without rhexis. It may show (A) two aspects: one clear band with thin irregular Z lines, not visible at histology (left-top; EM X 13,500) and thin (left-down; EM x 5,050) or dark increased electron density (right lower; E.M. x 4,400) visible at histology (PTHA x 640). In both pancellular and paradiscal lesions note the sarcolemma infolding corresponding to hypercontract sarcomeres and "squeezing" of mitochondria by the latter. The paradiscal lesion is present five minutes after the experimental intravenous infusion of catecholamines. (B) Wavy normal myocardial cells around a "hypercontraction center" (H & E x 250; EM x 5,200).

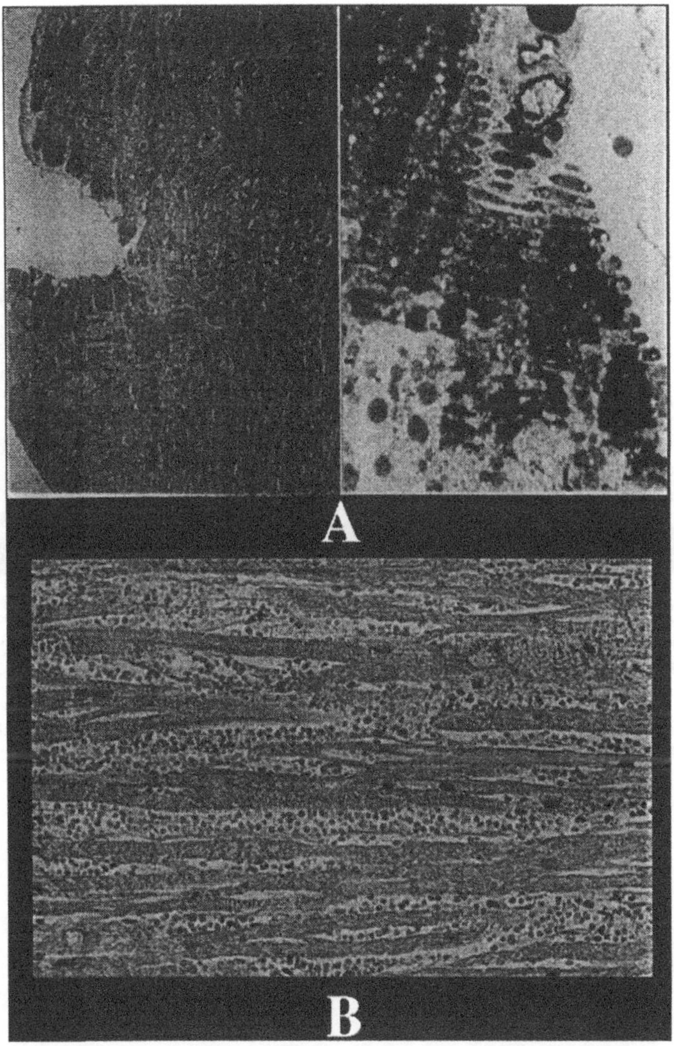

Figure 13. Other contraction band lesions: (A) Artifactual "cutting edge" lesion in an excised heart at transplantation. This lesion involves from two to five millimeters thick myocardial layer related to the cut edge. It shows hypercontracted sarcomeres with thickened Z line without rhexis (EM x 5, 200) and (H & E x 250). (B) Extensive pathological bands associated with massive interstitial hemorrhage found in reperfusion necrosis ("concentric hemorrhagic necrosis). Hemorrhage is not normally associated with CBN.

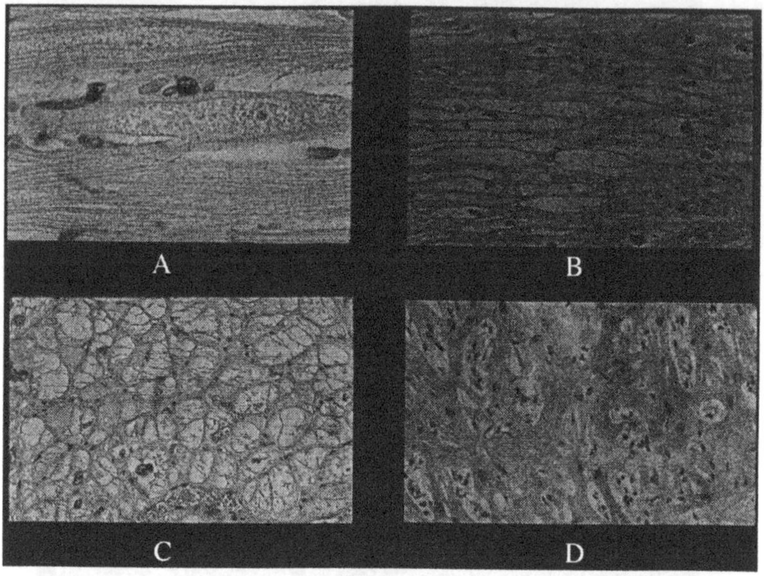


Figure 14. Colliquative myocytolysis. (A) Intramyocellular edema and small granules (mitochondria) with disappearance of myofibrils (H & E x 1000). (B) Longitudinal (H & E x 250) and (C) transverse sections (H &E x 40) showing a progressive "vacuolization" till empty sarcolemmal tubes remain. Note absence of any cellular reaction. This is the typical lesion of the heart in congestive heart failure, irrespective of its cause and is independent of coronary blood flow (D) Myocardial disarray represents another form of asynergic myocardium (see text).

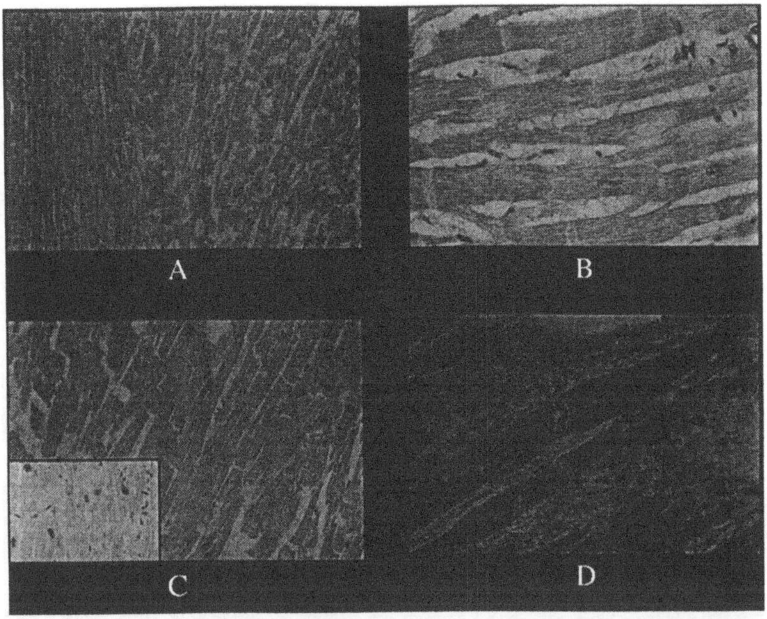


Figure 15. Morphologic findings in ventricular fibrillation. (A) Alternating myocardial bundles distended (left) and contracted plus segmentation (right; H & E x 100). (B) Stretched discs between contracted myocardial cells (H & E x 400). (C) Segmented myocardium (H & E x 400) with square instead of ovoidal nuclei because of contracted myocells (inlet, H & E x 400). (D) Stretching of sarcomeres in a myocardial cell between two contracted ones.

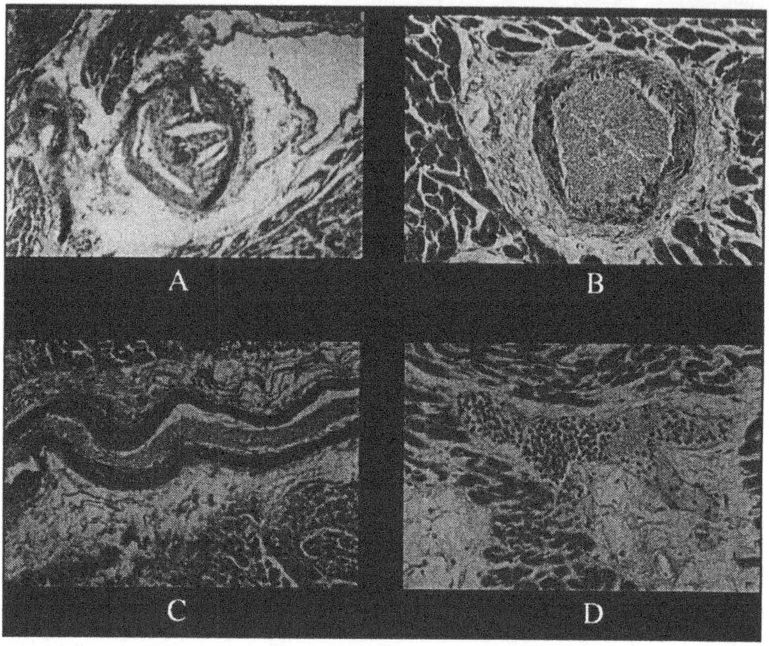


Figure 16. Intramural embolism and platelet aggregation in coronary heart disease. (A) Atheromatous embolus in an intramyocardial artery of a sudden/unexpected death case (H & E x 150). This embolus was found in normal myocardium and was the only one observed in hundreds of systematically studied hearts with and without coronary heart disease. (B) Arteriole filled by platelets in a normal subjects dead by accident (H & E x 180). (C) and (D) Layering of red cells, fibrin, polymorphonuclear leukocytes and platelets (H & E x 100) in terminal stasis. Changes found in a coagulum and not in a thrombus. There is no proof that platelet aggregates or fibrin-platelet thrombi or emboli are the cause of an acute coronary syndrome (see Fig. 23).

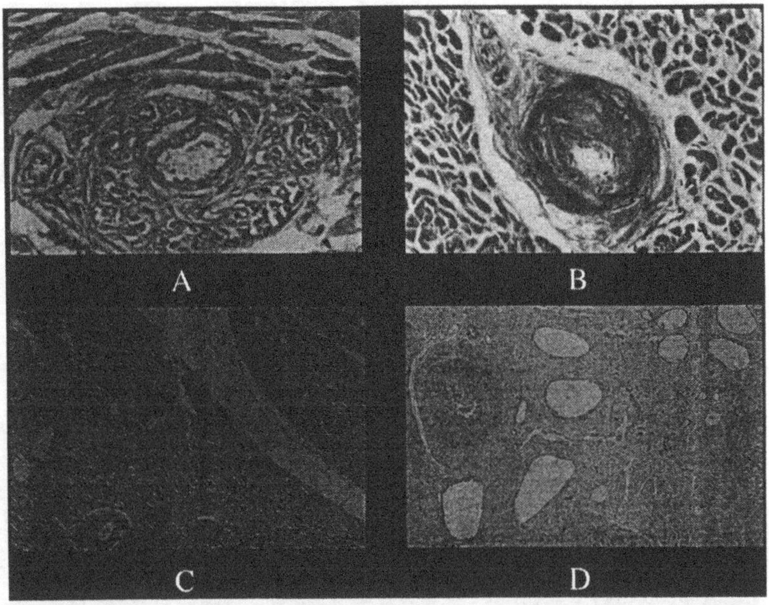


Figure 17. Medial hyperplasia obliterans. (A) Thickened media formed by circular and longitudinal bundles of smooth muscle cells (H & E x 150) leading (B) to progressive fibrosis (Mallory x 100). This change is seen in arterioles of the papillary muscles, trabeculae carneae and interventricular septum. Several vessels with this change (C) can be present in the normal myocardium of a normal subject (H & E x 100). (D) A similar pattern without prominent longitudinal bundles can be seen around scars (H & E x 100) and in areas of myocardial disarray. Note a vascular plexus formed by collaterals within the scar.

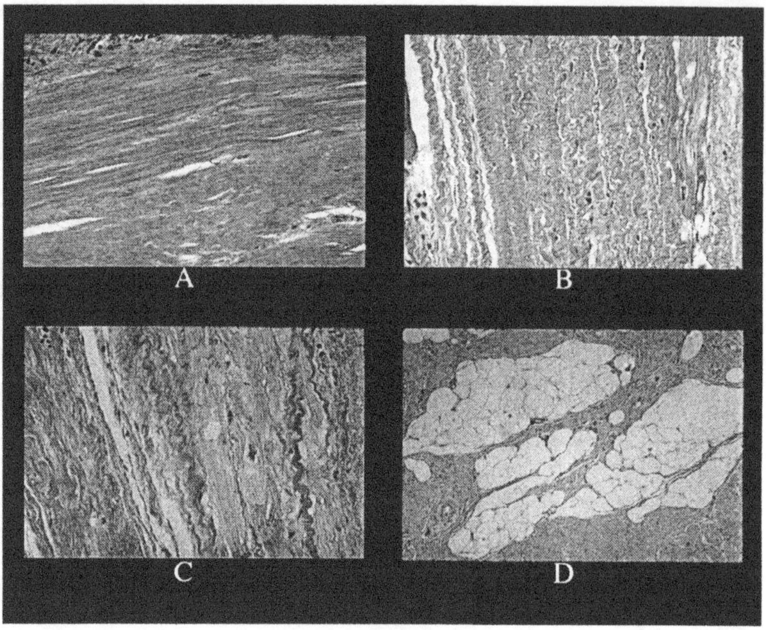

Figure 18. Myocardial fibrosis. This damage is considered by some to cause congestive heart failure. By measuring the size of myocardial fibrosis in percentage of the whole histological area examined in each heart in different conditions, it is clear that the amount of fibrosis, - even in coronary heart disease patients with resultant congestive heart failure - cannot explain the functional failure. With the exception of the central part of a healed infarct where there is dense collagen with straight, packed fibers (A; Gomori x 250), the collagen shows an undulating or wavy pattern (B; H & E x 250; C Gomory x 100). This indicates that any fibrous proliferation, including the collagen matrix, occurs after hypertrophy and in a beating myocardium. Fibrosis adapts its growth by assuming the wavy structure that cannot reduce contraction. (D) Any time collagen size is measured by a direct or indirect method, one must consider the frequent transformation of a scar into adipose tissue: (D) extensive lipomatous metaplasia in a huge myocardial scar (H & E x 250).

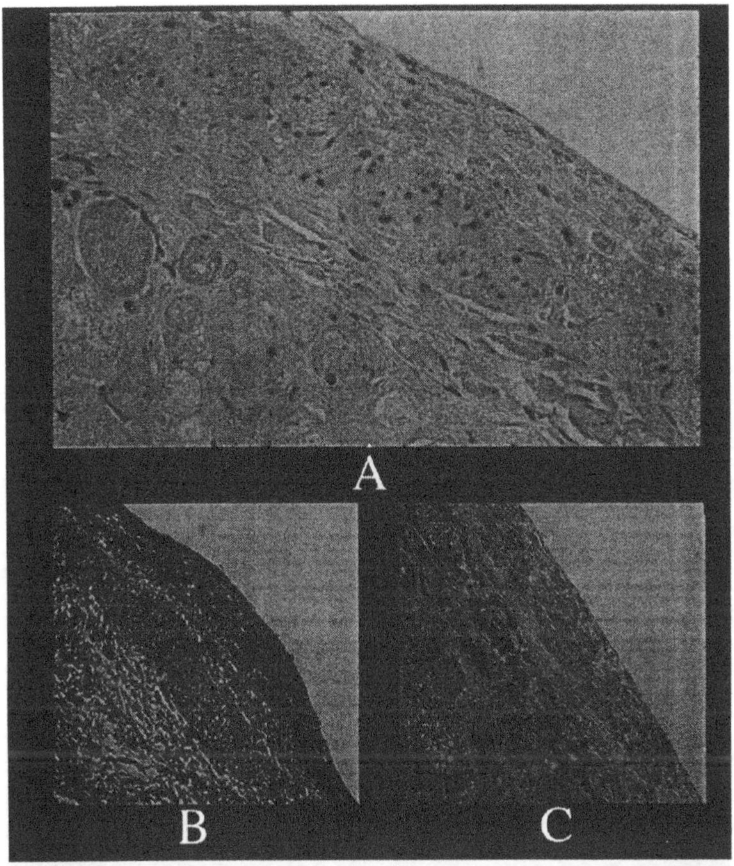


Figure 19. Endomyocardial myoelastofibrosis. Endocardial thickening is generally thought to result from the organization of endocardial thrombi. However in our CHF cases, we observed an endocardial thickening which starts as nodular smooth muscle cell hyperplasia (A; H & E x 250) followed by hyperplasia of elastic tissue (B; Weigert elastic x 100) and subsequent endocardial fibrosis (C; Mallory x 100).

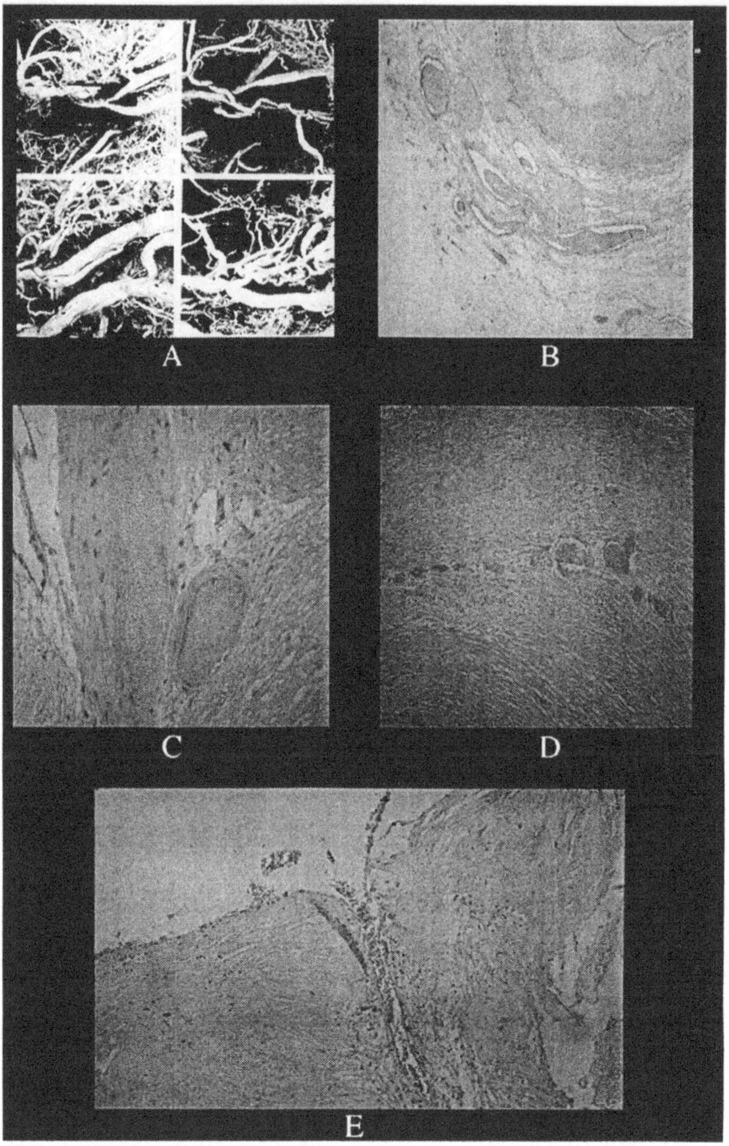


Figure 20. Vascularization of the coronary atherosclerotic plaque. (A) Plastic casts showing vessels bypassing a severe stenosis. This satellite system was documented histologically by serial section study of coronary artery plaques injected postmortem by radiopaque menstruum. It is formed by large, capillary-like adventitial vessels connected with proximal and distal secondary branches and (B) new formed arterioles with a well developed tunica media that are located in the atherosclerotic intima and (C) in turn are connected with intimal angiomatous plexuses (D) and the residual lumen (E; H&E x 250). Note that all vessels are arterial since the injected material penetrated within the arterial system only. In vivo, by cineangiography, these vessels can be partially seen and interpreted as possible imaging of an "active thrombogenic" plaque.

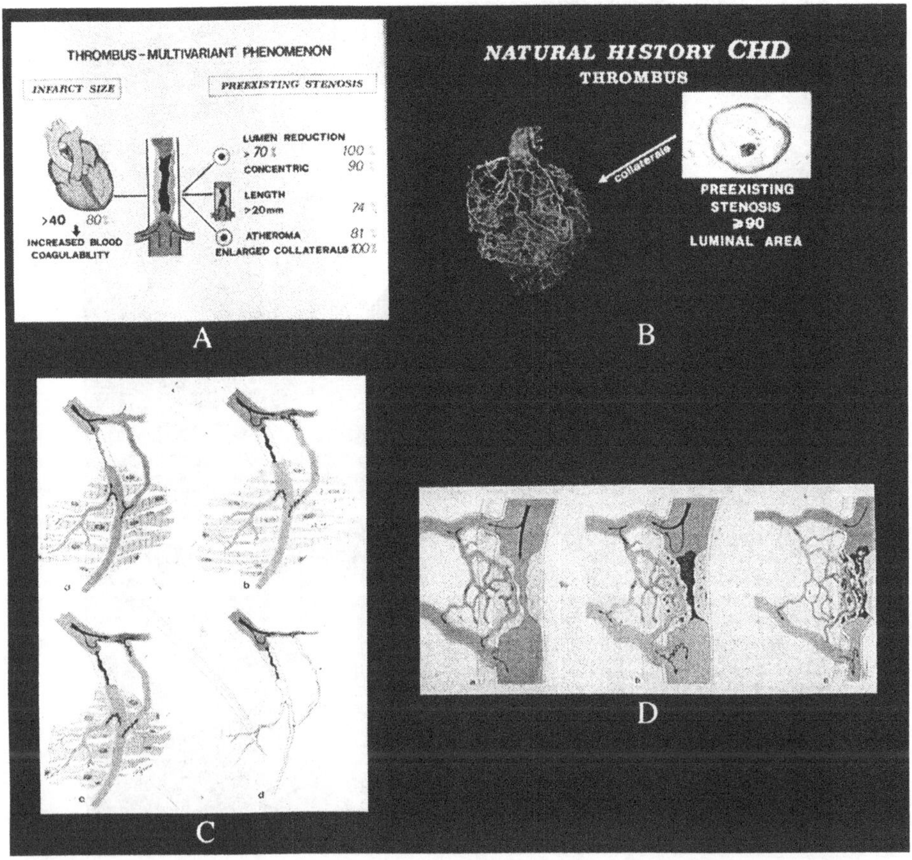


Figure 21. Secondary coronary thrombus formation. The thrombus is a multivariant phenomenon (A). Any time if forms it is located in a severe stenosis already bypassed by collaterals. It seems unrealistic to assume that the occlusion of a pin-point stenosis bypassed by collaterals (B) is the cause of a biological disaster i.e., an infarct. On the contrary, anytime there is an increased peripheral resistance resulting from myocardial asynergy and/or a still not clearly proven spasm, flow hindrance occurs at the plaque level bypassed by collaterals (C) with all the secondary changes within the plaque, i.e., hemorrhage, rupture, thrombus (D).

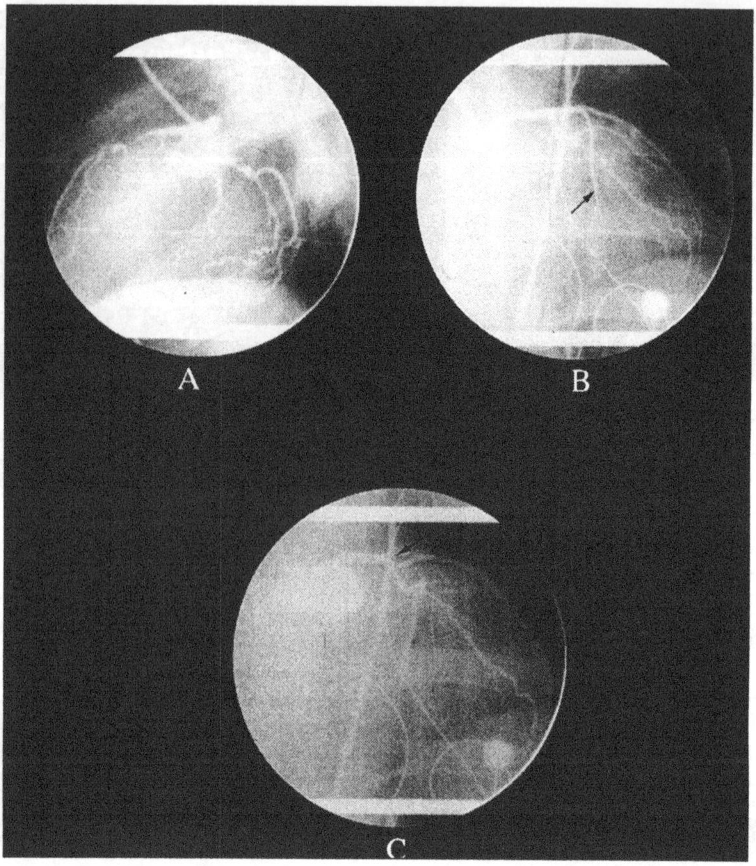


Figure 22. Angiographic coronary occlusion versus pseudocclusion. In a unique case of unstable angina monitored before and during the course of infarct formation and emergency therapies. In the few short periods of reperfusion, it was documented that the angiographic occlusion was in reality a blockage of flow due to increased intramyocardial resistance, likely secondary to extravascular compression by atonic myocardium stretched by the intraventricular pressure. (A) Severe angiographic stenosis of the left anterior descending branch associated with hypokinesis (not shown) of the depending myocardium; angiographic view before ECG ischemic changes. (B) ascending occlusion from distal to the origin of the artery (C). The heart excised at transplantation for irreversible congestive heart failure 12 months after the infarct, confirmed that there was no occlusion related to the infarct at any extramural and intramural level (see text).

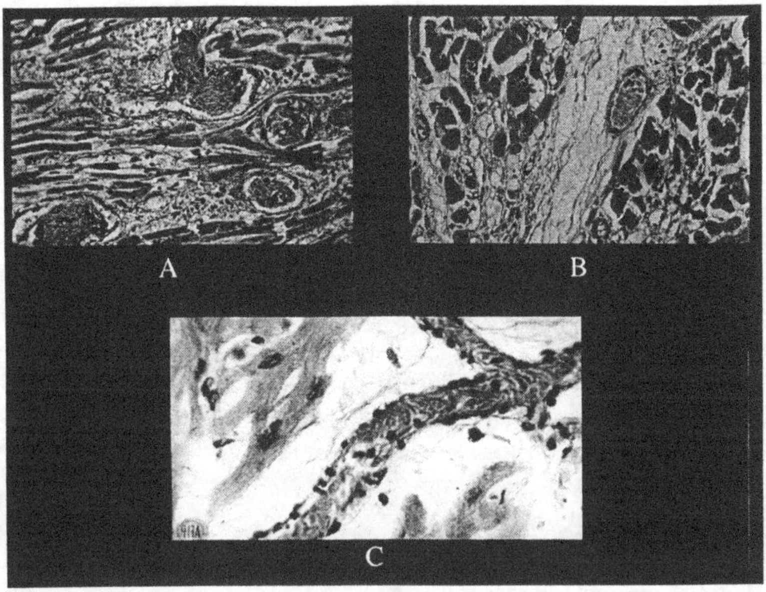


Figure 23. Small vessel diseases without evidence of coronary heart disease. (A) Thrombotic thrombocytopenic purpura, showing a diffuse, severe obstructive lesion of most intramural arterioles (Movat x 250) and platelet aggregates in most normal ones (B; Movat x 250) all without evidence of myocardial ischemic changes. In not one of the 39 cases personally examined was there a history, symptoms or signs of coronary heart disease. Only foci of nonhemorrhagic CBN necrosis not related to an occluded vessel, were found. (C) Similarly, we were unable to document any clinical and pathological finding of CHD in 52 sickle cell anemia patients, despite the entanglement (C) of sickle erythrocytes within vessels.

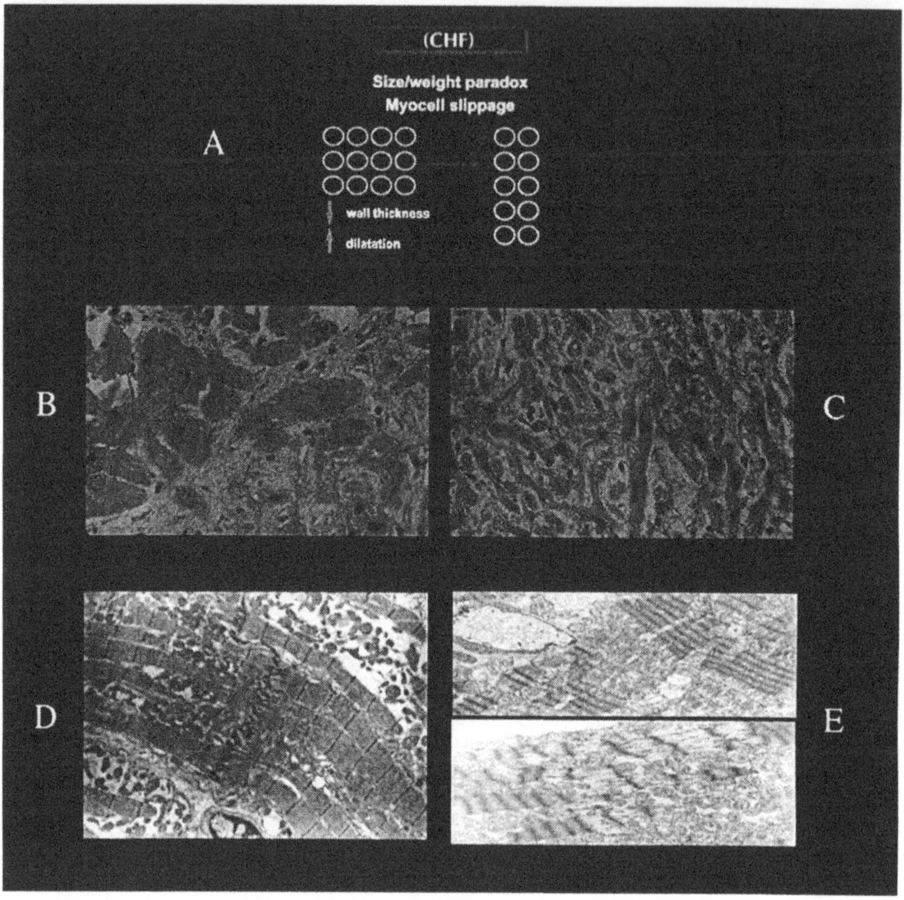


Figure 24. Myocardial stretching, slippage and new myocell formation have been proposed to explain the size/weight paradox (heavy hearts with normal wall thickness and normal myocardial cell size) in congestive heart failure. Interpenetration of the myocardial cells, i.e., slippage (A) contrasts with the normal myobridges between myocardial cells and collagen connection between the myocells and their surrounding. Furthermore, slippage should imply a destruction of all elements, blood vessels, lymphatic, nerves located in the interstitium. Similarly the hypothesis of myocardial cell growth contrasts with any proof that this happens in a beating myocardium. In an unique instance in an endocardial biopsy at the site of a previous one in a transplanted heart, we observed a node of myocardial cells structured like an atrio-ventricular node (C; H & E x 250) surrounded by normal sized myocardial cell (B; H & E x 250). A pattern never seen in all hearts we examined. The other explanation of stretching of myocellular elements is contradicted by histologic and ultrastructural findings. (C) normal contracted myocardial cells in a case of dilated cardiomyopathy (EM x 15,500) versus stretching (E).

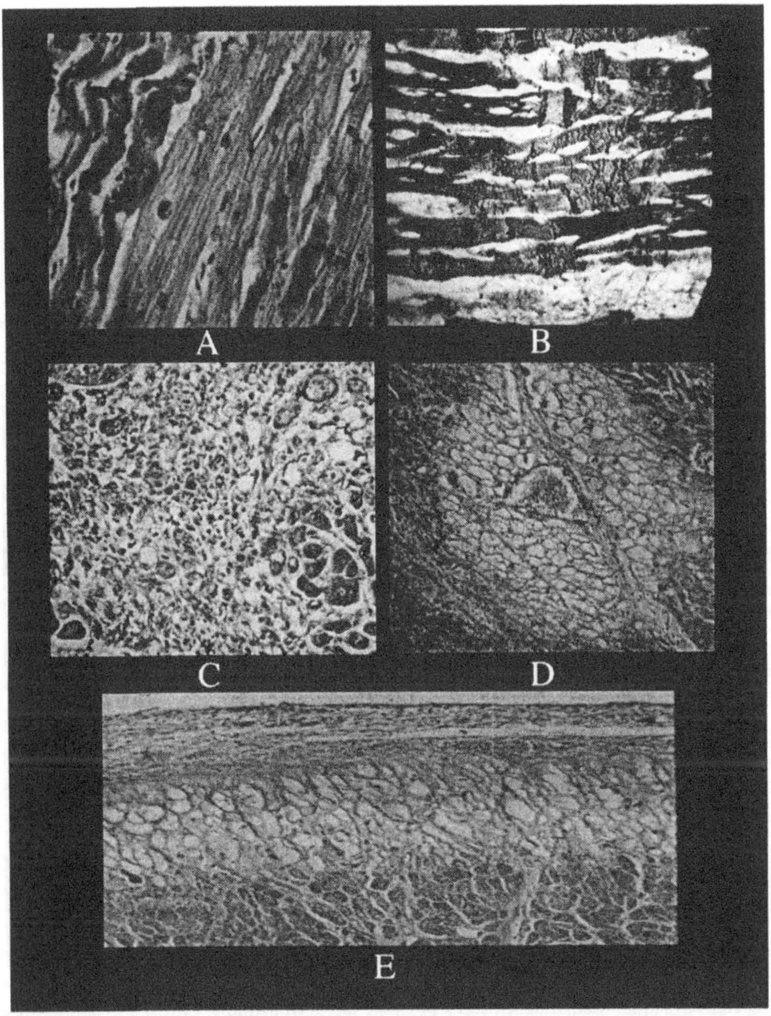


Figure 25. Association of different forms of myonecrosis in coronary heart disease. (A) Early eosinophilic infarct well defined from normal myocardium in which a few isolated myocardial cells show pathological contraction bands without hemorrhage (H & E x 250). (B) Extensive CBN around an acute infarct (PTHA x 100). (C) Alveolar healing stage of CBN associated with a recent infarct (H &E x 100). (D) Colliquative myocytolysis in the perivascular (H & E x 100) and (E) subendocardial myocardial regions preserved by infarct necrosis.

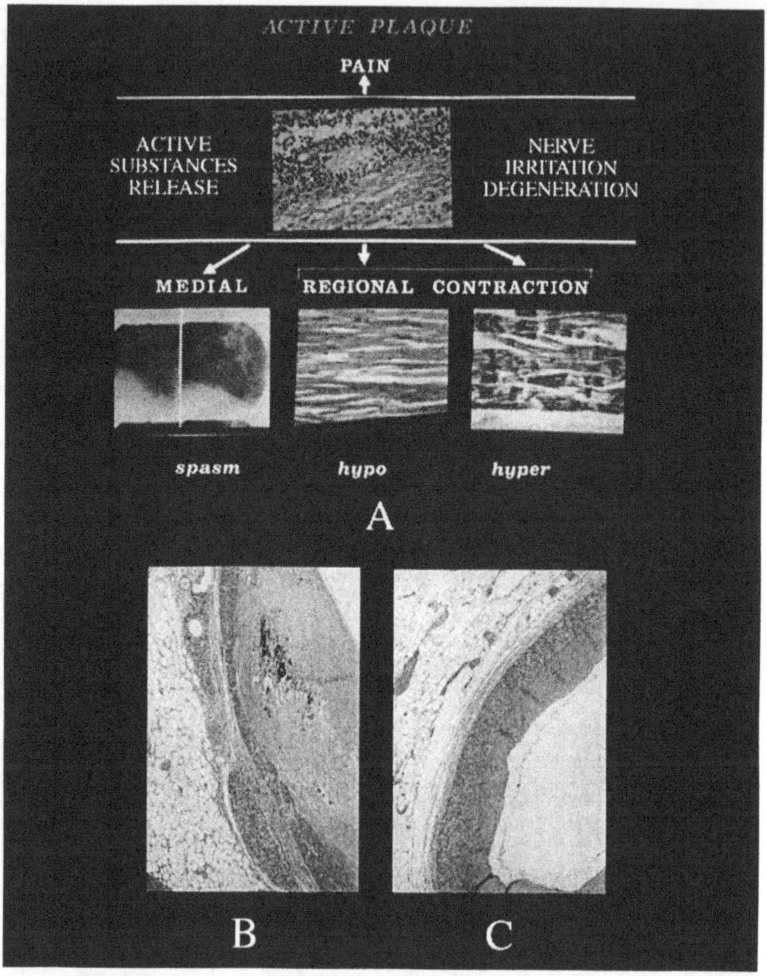


Figure 26. Active plaque. Medial neuritis as a possibile factor of intramyocardial adrenergic stress (A). A small plaques—usually undetected by angiography—shows this medial neuritis closely related to the atherosclerotic process (B; H & E x 100) while the normal wall of the same vessel (C; H & E x 100) is free of cellular infiltrates.

# TABLES

***Table 1. Variation of maximal and average diameter of the coronary collateral channels in different states in relation to anastomotic index***

| Status | Diameter µm Maximum | Average | Anastomotic Index Mean |
|---|---|---|---|
| Normal | 280 (180-350) | 200 (150-280) | 5 (3-6) |
| Atrophy | 240 (150-390) | 170 (100-250) | 4 (2-6) |
| Atrophy + hypoxia | 315 (260-400) | 218 (180-280) | 6 (5-12) |
| Hypertrophy | 350 (200-500) | 221 (130-350) | 7 (5-14) |
| Hypertrophy + hypoxia | 486 (300-700) | 304 (180-400) | 12 (5-19) |
| Hypoxia | 395 (299-395) | 249 (180-330) | 12 (9-19) |
| Mild (<69%) coronary stenosis | 320 (250-425) | 209 (165-225) | 7 (6-9) |
| + atrophy | 325 | 225 | |
| + hypertrophy | 318 (250-390) | 189 (150-250) | 8 (5-12) |
| + hypoxia | 400 (300-600) | 242 (200-400) | 12 (10-14) |
| Severe (70-99%) coronary stenosis | 345 (260-400) | 170 (180-200) | 8 (7-10) |
| + atrophy | 450 (300-600) | 150 (140-160) | 11 (9-12) |
| + hypertrophy | 428 (230-1900) | 254 (150-460) | 11 (5-22) |
| + hypoxia | 461 (300-1000) | 307 (150-500) | 14 (5-25) |
| Coronary occlusion | 685 (350-1250) | 347 (200-450) | 16 (5-33) |
| + atrophy | 572 (500-650) | 350 (300-400) | 15 (11-18) |
| + hypertrophy | 729 (290-1690) | 413 (125-600) | 19 (5-35) |
| + hypoxia | 512 (250-780) | 292 (170-550) | 14 (6-20) |
| Multiple coronary occlusions | 780 (288-2000) | 467 (100-600) | 22 (15-38) |

Maximun diameter=largest collateral found in single case.
Average diameter=average diameter of the larger collaterals (> 100µm) found in single cases.
Anastomotic index see text.

***Table 2. Correlative aging between occlusive thrombus and myocardial infarction***

| Thrombus | Age | Infarct Necrosis |
|---|---|---|
| Fibrin/platelets | 6-8 hours | Normal myocardium or PMN infiltration |
| Endothelial sprouting/capillary ingrowth | 7 days | Initial macrophagic reaction at periphery. Very early fibrosis |
| Early organization/collagen fibril deposition | 15 days | Diffuse macrophages/alveolar appearance |
| Early to old organization | > 20 days | Recent to old fibrosis |

***Table 3. Maximal lumen diameter reduction and number of vessels with severe (≥ 70%) stenosis in acute infarct (AMI) cases without and with postmortem fixation of coronary arteries under pressure****

| AMICases | Maximal Lumen Reduction% | | | | No. of Vessels ≥ 70 Stenosis | | |
|---|---|---|---|---|---|---|---|
| | ≤ 69 | 70 | 80 | ≥ 90 | 1 | 2 | 3 |
| 100 without | 11 | 20 | 27 | 42 | 38 | 36 | 15 |
| 100 with pressure | 6 | 18 | 29 | 47 | 39 | 35 | 20 |
| Total | 17 | 38 | 56 | 89 | 77 | 71 | 35 |

* see text for details

**Table 4. *Frequency distribution of coronary atherosclerotic stenosing plaques in 97 adult "normal" subjects (88 men - 9 women) dying by accident, without history and postmortem finding of any disease***

| | | Maximal Lumen Reduction %* | | | | | | Stenosis vs No. Vessels | | | | | |
|---|---|---|---|---|---|---|---|---|---|---|---|---|---|
| | | | | | | | | Anytype | | | Severe (≥ 70%) | | |
| Age | Cases | 0 | < 50 | 50-69 | 70-79 | 80-89 | > 90 | 1 | 2 | 3 | 1 | 2 | 3 or more |
| < 39 | 10 | 2 | 3 | 4 | 1 | - | - | 3 | 3 | 2 | - | 1 | - |
| 40-49 | 13 | 2 | 6 | 2 | 2 | 1 | - | 1 | 5 | 5 | 2 | 1 | - |
| 50-59 | 28 | 4 | 7 | 7 | 5 | 4 | 1 | 3 | 6 | 15 | 6 | 3 | 1 |
| 60-69 | 23 | - | 2 | 9 | 5 | 4 | 3 | 5 | 3 | 15 | 7 | 4 | 1 |
| > 70 | 23 | - | 2 | 9 | 6 | 4 | 2 | - | 4 | 19 | 7 | 4 | 1 |
| Total | 97 | 8 | 20 | 31 | 19 | 13 | 6 | 12 | 21 | 56 | 22 | 13 | 3 |

* % lumen diameter

**Table 5. *Percentage distribution of morphological variables of atherosclerotic and fibrous plaques in relation to lumen and intimal thickness***

| Lumen Reduction (%) | Total Section | Absent | Coronary Atherosclerotic Plaque Variables P | AT | IV | HR | CA | ILI | ALI | ALI+ILI | FP |
|---|---|---|---|---|---|---|---|---|---|---|---|
| 0 | **1.519** | 97 | 0.5 | 0.3 | 0.4 | 0.2 | 0.2 | 0.3 | 0.5 | 0.9 | - |
| < 50 | **687** | 34 | 52 | 27 | 31 | 3 | 27 | 26 | 20 | 32 | 45 |
| 50-69 | **632** | 8 | 70 | 54 | 63 | 8 | 48 | 52 | 43 | 62 | 20 |
| 70-79 | **298** | 3 | 71 | 75 | 74 | 19 | 57 | 67 | 54 | 74 | 12 |
| 80-89 | **262** | 1 | 71 | 77 | 82 | 30 | 63 | 69 | 64 | 84 | 9 |
| ≥ 90 | **242** | 2 | 58 | 74 | 78 | 33 | 69 | 69 | 66 | 82 | 16 |
| **Intimal thickness[c] (μm)** | | | | | | | | | | | |
| ≤ 299 | **162** | 93 | 3 | 2 | 3 | - | 3 | 2 | 1 | 2 | 96 |
| 300-599 | **199** | 34 | 50 | 16 | 24 | 1 | 21 | 23 | 14 | 28 | 49 |
| 600-999 | **544** | 8 | 69 | 42 | 57 | 6 | 38 | 45 | 36 | 54 | 24 |
| 1000-1.999 | **926** | 3 | 75 | 69 | 75 | 18 | 60 | 62 | 53 | 73 | 12 |
| ≥ 2000 | **290** | 4 | 57 | 76 | 63 | 30 | 64 | 68 | 62 | 79 | 13 |

Abbreviations: Absent, no morphologic variables; ALI, adventitial lymphocytic infiltration; AT, atheroma; P, proteoglycans; CA, calcification; FP, fibrous plaque without atheroma and/or basophilia; HR, intimal hemorrhage; ILI, intimal lymphocytic infiltration; IV, intimal vascularization.
[c] Only sections with lumen reduction considered here.

*Table 6.* ***Myocardial asynergy versus structural damage***

| Authors | | Disease | Method of Study | Cases No. | Segments No. | False-Positive Normal Myoc. | | | False-Negative Myoc. Necrosis | | |
|---|---|---|---|---|---|---|---|---|---|---|---|
| | | | | | | Hypokinesis | | Akin/Dyskin | Transmural | ? | Non Trans. |
| Cabin et al | 1987 | Acute infarct | RA/A | 23 | 228 | 27(12%) | | 6(3%) | 11(10%) | | 34(15%) |
| Sinusas et al | 1984 | Chronic IHD (48)<br>Valvular Cpt (3)<br>Dilated Cmp (4) | RA/A | 55 | 372 | - | 35(9%) | - | - | 61(16%) | - |
| Ideker et al | 1978 | Chronic IHD | Vq/A | 24 | 72 | - | 8(11%) | - | - | 2(3%) | - |
| Hutchins et al | 1977 | Chronic IHD (24)<br>Valvular Cpt (4) | V/A | 28 | 140 | 56(40%) | | 16(11%) | - | | 1(0.7%) |
| Stinson et al | 1977 | Chronic IHD | V/B | 110 | 110 | 12(11%) | | 4(4%) | - | | 15(14%) |
| Bodenheimer et al | 1976 | Chronic IHD | V/B | 25 | 29 | 11(38%) | | - | - | | - |

RA, radionucleide angiography; V, ventriculography; q, quantitative; A, autopsy; B, biopsy; Cmp, cardiomyopathy; Cpt, cardiopathy; IHD, ischemic heart disease.
False-positive: asinergy + normal myocardium
False-negative: myocardial necrosis + normal contraction

**Table 7. Sex and age distribution of cases studied personally**

| | | Age (Years) | | | | | |
|---|---|---|---|---|---|---|---|
| | | < 40 | 40-49 | 50-59 | 60-69 | ≥70 | Total |
| SD | Men | 30 | 31 | 63* | 41 | 17 | 182 |
| | Women | 5 | 2 | 5 | 4 | 10* | 26 |
| | **Total** | **35** | **33** | **68** | **45** | **27** | **208** |
| AMI | Men | 4 | 18 | 35 | 45+ | 36 | 138 |
| | Women | - | 4 | 9 | 22 | 27+ | 62 |
| | **Total** | **4** | **22** | **44** | **67** | **63** | **200** |
| AD | Men | 9 | 11 | 27 | 20 | 21 | 88 |
| | Women | 1 | 2 | 1 | 3 | 2 | 9 |
| | **Total** | **10** | **13** | **28** | **23** | **23** | **97** |

Man/Woman ratio: SD = 7.0; AMI = 2.2 ; AD = 9.7
SD, sudden/unexpected coronary death; AMI acute myocardial infarct; AD, accidental death in normal people

**Table 8. Frequency of maximal lumen reduction and number of vessels with severe stenosis (≥ 70% lumen diameter) in relation to age in sudden/unexpected coronary death (SD), acute myocardial infarct (AMI) and in healthy subjects dying by accident (AD)**

| Age (Years) | Lumen reduction (%) < 69 | 70-79 | 80-89 | > 90 | Total | Stenosis ≥ 70% in 1 | 2 | 3 +Vess | Total |
|---|---|---|---|---|---|---|---|---|---|
| ***≤ 39*** | | | | | | | | | |
| SD | 18* | 1 | 9 | 7 | 35 | 7 | 6 | 4 | 17 |
| AMI | 1 | 2 | - | 1 | 4 | 2 | - | 1 | 3 |
| AD | 9 | 1 | - | - | 10 | - | 1 | - | 1 |
| ***40-69*** | | | | | | | | | |
| SD | 28 | 20 | 39 | 59* | 146 | 37 | 45 | 36 | 118 |
| AMI | 13 | 21 | 37 | 62* | 133 | 46 | 50 | 24 | 120 |
| AD | 39* | 12 | 9 | 4 | 64 | 14 | 9 | 2 | 25 |
| ***≥ 70*** | | | | | | | | | |
| SD | 5 | 8 | 5 | 9* | 27 | 9 | 9 | 4 | 22 |
| AMI | 3 | 15 | 19 | 26* | 63 | 29 | 21 | 10 | 60 |
| AD | 11 | 6 | 4 | 2 | 23 | 8 | 3 | 1 | 12 |
| ***Total*** | | | | | | | | | |
| SD | 51 | 29 | 53 | 75* | 208 | 53 | 60 | 44 | 157 |
| AMI | 17 | 38 | 56 | 89* | 200 | 77* | 71 | 35 | 183 |
| AD | 59* | 19 | 13 | 6 | 97 | 22 | 13 | 3 | 38 |

* in excess $P < 0.01$

***Table 9. Maximal lumen reduction and number of main subepicardial coronary arteries with severe (> 70% lumen diameter) stenosis***

| Source | No. | % Maximal Lumen Reduction | | | | | | Severe Stenosis in | | |
|---|---|---|---|---|---|---|---|---|---|---|
| | Cases | 0 | < 50 | 50-69 | 70-79 | 80-89 | > 90 | 1 | 2 | 3 Vess |
| AMI Ist | 145 | 3 (2) | 3 (2) | 10 (7) | 30 (21) | 45 (31) | 54 (37) | 61 (42) | 49 (34) | 19 (13) |
| AMI chr | 55 | - | 1 (2) | - | 8 (14) | 11 (20) | 35 (64) | 16 (29) | 22 (40) | 16 (29) |
| Total | 200 | 3 (1) | 4 (2) | 10 (5) | 38 (19) | 56 (28) | 89 (44) | 77(38) | 71 (35) | 35 (17) |
| SD Ist | 133 | 10 (8) | 18 (13) | 18 (13) | 21 (16) | 39 (29) | 27 (20) | 40 (30) | 34 (25) | 13 (10) |
| SD chr | 75 | - | - | 5 (7) | 8 (11) | 14 (18) | 48 (64) | 13 (17) | 26 (35) | 31 (41) |
| Total | 208 | 10 (5) | 18 (9) | 23 (11) | 29 (14) | 53 (25) | 75 (36) | 53 (25) | 60 (29) | 44 (21) |
| NCA | 100 | 7 (5) | 10 (10) | 17 (17) | 11 (11) | 24 (24) | 31 (31) | 26 (26) | 18 (18) | 22 (22) |
| AD | 97 | 8 (8) | 20 (21) | 31 (32) | 19 (19) | 13 (14) | 6 (6) | 22 (23) | 13 (13) | 3 (3) |

AMI, acute myocardial infarct; SD, sudden/unexpected coronary death; NCA, non cardiac atherosclerotic patients; AD, normal subjects dying from accident; Ist = first episode and chr=chronic, previous episodes of CHD. In parenthesis (%).

***Table 10. Percentage distribution of morphological variables and "fibrous plaque" in different groups in relation to lumen reduction and intimal thickness[a]***

| Group (Sections Studied) | Absent | FP[a] | P | AT | IV | HR | CA | ILI | ALI | ILI+ALI |
|---|---|---|---|---|---|---|---|---|---|---|
| | Morphological Variables | | | | | | | | | |
| ***No lumen reduction*** | | | | | | | | | | |
| AMI (309) | 96 | - | 2 | 1 | 2 | 1 | 1 | 1 | 2 | 3 |
| CI (225) | 99 | - | 0.4 | - | - | - | 0.4 | 1 | - | 1 |
| SDF (245) | 99 | - | 0.4 | - | 0.4 | - | - | 0.4 | - | 0.4 |
| SDNF (335) | 99 | - | 0.3 | - | - | - | 0.3 | - | 1 | 1 |
| AD (405) | 100 | - | - | - | - | - | - | - | - | - |
| **Total (1.519)** | **99** | **-** | **0.5** | **0.3** | **0.4** | **0.2** | **0.3** | **0.5** | **0.5** | **1** |
| ***Lumen reduction ≤ 69%*** | | | | | | | | | | |
| AMI (264) | 9 | 17 | 72 | 56 | 43 | 16 | 50 | 55 | 46 | 67 |
| CI (77) | 23 | 43 | 40 | 43 | 40 | 9 | 43 | 51 | 40 | 57 |
| SDF (346) | 25 | 40 | 54 | 34 | 48 | 2 | 35 | 37 | 26 | 43 |
| SDNF (325) | 22 | 32 | 65 | 41 | 53 | 4 | 36 | 34 | 31 | 42 |
| AD (307) | 28 | 37 | 59 | 30 | 41 | 2 | 27 | 28 | 20 | 35 |
| ***Lumen reduction ≥ 70%*** | | | | | | | | | | |
| AMI (227) | 0.4 | 11 | 60 | 78 | 72 | 43 | 74 | 74 | 65 | 85 |
| CI (98) | 4 | 14 | 44 | 73 | 68 | 32 | 63 | 76 | 67 | 87 |
| SDF (225) | 3 | 12 | 76 | 72 | 82 | 21 | 58 | 64 | 65 | 77 |
| SDNF (188) | 2 | 13 | 72 | 76 | 86 | 20 | 61 | 64 | 54 | 75 |
| AD (64) | 3 | 8 | 80 | 75 | 72 | 8 | 44 | 64 | 42 | 72 |
| **Grand total (2.121)** | **14** | **25** | **63** | **53** | **58** | **14** | **47** | **50** | **42** | **59** |
| ***Intimal thickness (μm) ≤ 599*** | | | | | | | | | | |
| AMI (43) | 26 | 39 | 58 | 21 | 28 | 5 | 21 | 39 | 21 | 44 |
| CI (3) | 33 | 67 | 33 | 33 | 66 | - | - | 67 | 33 | 67 |
| SDF (105) | 69 | 80 | 19 | 9 | 10 | - | 9.5 | 11 | 4 | 13 |
| SDNF (98) | 59 | 72 | 26 | 9 | 13 | - | 16 | 10 | 9 | 13 |
| AD (112) | 66 | 71 | 28 | 6 | 12 | - | 11 | 6 | 6 | 9 |
| ***Intimal thickness (μm) ≥ 600*** | | | | | | | | | | |
| AMI (448) | 3 | 12 | 67 | 71 | 59 | 31 | 65 | 66 | 58 | 78 |
| CI (172) | 12 | 26 | 42 | 60 | 56 | 22 | 55 | 65 | 56 | 74 |
| SDF (466) | 4 | 18 | 72 | 58 | 73 | 11 | 52 | 56 | 50 | 66 |
| SDNF (415) | 4 | 14 | 78 | 65 | 78 | 12 | 52 | 53 | 47 | 64 |
| AD (259) | 5 | 15 | 77 | 52 | 61 | 4 | 39 | 46 | 32 | 55 |

[a]Only sections with lumen reduction considered here.
[b]Abbreviations: AD, accidental death; ALI, adventitial lymphocytic infiltration; AMI, acute myocardial infarction; AT, atheroma; P, proteoglycan; CA, calcification; CI, chronic ischemia; FP, fibrous plaque without atheroma and/or basophilia; HR, intimal hemorrhage; ILI, intimal lymphocytic infiltration; IV, intimal vascularization; SDNF, sudden/unexpected death without, SDF, sudden/unexpected death with monofocal extensive myocardial fibrosis (≥ 10% of the left ventricular mass).

***Table 11. Lympho-plasmacellular inflammatory reaction (medial neuritis) in coronary atherosclerotic plaques in different groups of population***

| | Cases | MN% | Stenoses | MN% | Mild | Moder | Severe |
|---|---|---|---|---|---|---|---|
| Acute infarct | 100 | 100 | 491 | 75 | 34 | 21 | 20 |
| Chronic IHD | 50 | 88 | 175 | 74 | 32 | 13 | 29 |
| Sudden/unex. death | 208 | 83 | 1084 | 55 | 29 | 16 | 10 |
| Healthy controls | 97 | 64 | 371 | 41 | 22 | 14 | 5 |

MN=medial neuritis; *total stenoses of any grade.

***Table 12. Percentage distribution of lympho-plasmacellular inflammatory reaction (IR), proteoglycan accumulation (PA) and atheroma (AT) related to intimal thickness and lumen reduction***

| Intimal Thickness µm | IR | PA | AT | Lumen Reduction % | IR | PA | AT |
|---|---|---|---|---|---|---|---|
| 300 | 2 | 3 | 2 | < 50 | 32 | 52 | 27 |
| 600 | 28 | 50 | 16 | 50-69 | 62 | 70 | 54 |
| 1000 | 54 | 69 | 43 | 70-79 | 74 | 71 | 75 |
| 2000 | 73 | 75 | 69 | 80-89 | 84 | 71 | 77 |
| > 2000 | 79 | 57 | 76 | > 90 | 82 | 58 | 74 |

***Table 13. Main significant variations of plaque variables in different groups***

| | AMI | CI | SD1st | SDCH | AD |
|---|---|---|---|---|---|
| All stenoses | + | - | + | ns | - |
| Severe stenosis (≥ 70 %) | + | + | ns | ns | - |
| Concentric plaque | + | ns | - | ns | ns |
| Semilunar plaque | - | ns | + | ns | ns |
| Short stenoses (3mm) | + | ns | ns | ns | + |
| Long stenoses (30 mm) | - | ns | ns | ns | ns |
| Intimal thickness ≥ 2000 μm | + | + | ns | ns | - |
| Intimal thickness ≤299 μm | - | - | - | - | + |
| Medial thickness ≥200 μm | ns | + | ns | ns | ns |
| Medial thickness ≤99 μm | + | ns | + | ns | ns |
| Atheroma or advential lymphocytic infiltrate or intimal lymphocytic infiltrate or calcification | + | + | ns | ns | - |
| Same variables in most stenoses in single case | + | + | ns | ns | - |
| Intimal hemorrhage | + | ns | - | - | - |
| Proteoglycans | + | - | ns | + | + |
| Intima vascularization | - | - | + | + | ns |
| Fibrosis plaque | - | - | + | - | + |
| Thrombosis | | | | | |
| Acute occlusive | + | ns | ns | ns | - |
| Acute mural | + | ns | ns | ns | - |
| Old occlusive | ns | + | ns | ns | - |
| Old mural | ns | ns | ns | ns | - |

AD, accidental death; AMI, acute myocardial infarction; CI, chronic ischemia; SD1st, sudden death first episode; SDCH, sudden death in chronic CHD
+ in excess; - in deficit; ns in expected range.

***Table 14. Maximal lumen reduction and number of vessels with severe stenosis (≥ 70%) in relation to heart weight in different groups****

| Source Heart Weight (g) | Cases | Lumen Reduction % | | Stenosis ≥ 70% in | | |
|---|---|---|---|---|---|---|
| | | < 69 | ≥ 70 | 1 | 2 | ≥3 Vess |
| ***AMI 1st*** | | | | | | |
| < 500 | 43 | 6 | 37 | 16 | 17 | 4 |
| ≥ 500 | 27 | 4 | 23 | 13 | 6 | 4 |
| **Total** | **70** | **10** | **60** | **29** | **23** | **8** |
| ***AMI chr*** | | | | | | |
| < 500 | 14 | - | 14 | 3 | 7 | 4 |
| ≥ 500 | 16 | 1 | 15 | 6 | 6 | 3 |
| **Total** | **30** | **1** | **29** | **9** | **13** | **7** |
| ***SD 1st*** | | | | | | |
| < 500 | 76 | 24 | 52 | 29 | 21 | 2 |
| ≥ 500 | 57 | 22 | 35 | 11 | 13 | 11 |
| **Total** | **133** | **46** | **87** | **40** | **34** | **13** |
| ***SD chr*** | | | | | | |
| <500 | 18 | 2 | 16 | 3 | 6 | 7 |
| ≥ 500 | 57 | 3 | 54 | 10 | 20 | 24 |
| **Total** | **75** | **5** | **70** | **13** | **26** | **31** |
| ***AD*** | | | | | | |
| < 500 | 87 | 52 | 35 | 21 | 11 | 3 |
| ≥ 500 | 10 | 7 | 3 | 1 | 2 | - |
| **Total** | **97** | **59** | **38** | **22** | **13** | **3** |

* In AMI the relation between heart weight and lumen reduction has been calculated in 100 cases. AMI, acute myocardial infarct; SD, sudden/unexpected death; AD, accidental death in normal people; 1st, first episode of coronary heart disease; chr, chronic CHD.

***Table 15. Frequency of acute occlusive and mural thrombi in acute infarcts and sudden/unexpected coronary deaths in relation to extensive fibrosis***

| Source | Cases | Thrombus | |
|---|---|---|---|
| | | Occlusive | Mural |
| AMI 1st | 145 | 60 | 29 |
| AMI chr | 55 | 22 | 7 |
| Total | 200 | 82 | 36 |
| SD 1st | 133 | 11 | 14 |
| SD chr | 75 | 21 | 8 |
| Total | 208 | 32 | 22 |

In healthy control group only one mural thrombus found. For legends see Table 14.

***Table 16. Acute coronary thrombus in 200 acute infarcts (AMI) and 208 sudden/ unexpected coronary death (SD) in relation to atherosclerotic (ATS) plaque variables, infarct size and survival. Percentage distribution***

| | AMI | | SD | |
|---|---|---|---|---|
| **Thrombus** | **Occlusive** | **Mural** | **Occlusive** | **Mural** |
| Total | 41 | 18 | 15 | 11 |
| **ATS plaque stenosis** | | | | |
| ≤ 69 | 7 | 14 | - | 9 |
| 70-79 | 33 | 36 | 16 | 20 |
| 80-89 | 35 | 19 | 47 | 45 |
| > 90 | 24 | 31 | 38 | 32 |
| Length mm | | | | |
| ≤5 | 6 | 19 | 6 | 9 |
| 5-20 | 38 | 39 | 19 | 27 |
| > 20 | 56 | 42 | 75 | 64 |
| Type | | | | |
| concentric | 100 | 100 | 94 | 91 |
| atheromatous | 84 | 81 | 75 | 82 |
| **Medial neuritis** | 92 | 82 | 92 | 79 |
| **Infarct size** | | | | |
| ≤ 10 | 20 | 17 | - | - |
| 11-20 | 32 | 19 | - | - |
| 21-30 | 48 | 33 | - | - |
| 31-40 | 44 | 19 | - | - |
| 41-50 | 78 | 8 | - | - |
| > 50 | 86 | 3 | - | - |
| **Survival days** | | | | |
| ≤2 | 29 | - | - | - |
| 3-10 | 51 | - | - | - |
| 11-30 | 45 | - | - | - |
| **Survival minutes** | | | | |
| < 10 | - | - | 12 | - |
| 10-60 | - | - | 23 | - |
| 61-180 | - | - | 30 | - |

* in percent left ventricular/interventricular septum mass.

**Table 17. Relationship of infarct size (% left ventricular mass), location in the left ventricular wall and acute occlusive thrombus in supplying artery**

| Location in LV | Infarct Size (%) ≤ 10 | | 11-20 | | 21-30 | | 31-40 | | 41-50 | | > 50 | |
|---|---|---|---|---|---|---|---|---|---|---|---|---|
| | P | T | P | T | P | T | P | T | P | T | P | T |
| Inner 1/3 | 22 | 2 | 1 | 1* | - | - | - | - | - | - | - | - |
| Inner 2/3 | 22 | 6 | 18 | 7 | 14 | 7 | 4 | 2 | 1 | 1 | 3 | 3 |
| Transmural | 16 | 4 | 18 | 4 | 30 | 14 | 23 | 10 | 17 | 13 | 11 | 9 |
| Total | 60 | 12 | 37 | 11 | 44 | 21 | 27 | 12 | 18 | 14 | 14 | 12 |

P = patients; T = thrombi, LV=left ventricle

**Table 18. Distribution of infarct size (% left ventricular mass) in 200 consecutive acute infarct cases without (AMI 1st episode) and with (AMI chronic) extensive myocardial fibrosis**

| Source | Cases | Infarct Size (%) ≤ 10 | 11-20 | 21-30 | 31-40 | 41-50 | > 50 |
|---|---|---|---|---|---|---|---|
| AMI 1st | 145 | 32 | 30 | 32 | 24 | 14 | 13 |
| AMI chr | 55 | 28 | 7 | 12 | 3 | 4 | 1 |
| Total | 200 | 60 | 37 | 44 | 27 | 18 | 14 |

**Table 19. Infarct size, (% left ventricular mass) maximal lumen reduction and frequency of thrombus in supplying artery in 200 acute fatal myocardial infarcts**

| Infarct Size % | Cases | Luminal Stenosis (%) ≤ 59 | | | 60-69 | | | 70-79 | | | 80-89 | | | ≥ 90 | | |
|---|---|---|---|---|---|---|---|---|---|---|---|---|---|---|---|---|
| | | N | O | M | N | O | M | N | O | M | N | O | M | N | O | M |
| ≤ 10 | 60 | 3 | 1 | 1 | 5 | - | - | 10 | 1 | 2 | 8 | 7 | - | 16 | 3 | 3 |
| 11-20 | 37 | 4 | - | - | 1 | - | - | 5 | 6 | 3 | 4 | 3 | - | 4 | 3 | 4 |
| 21-30 | 44 | 2 | - | - | 1 | - | 2 | 2 | 7 | 3 | 3 | 7 | 4 | 3 | 7 | 3 |
| 31-40 | 27 | 3 | - | - | 1 | 1 | 1 | 3 | 5 | 3 | 2 | 3 | 2 | - | 2 | 1 |
| 41-50 | 18 | - | - | - | - | - | 1 | - | 3 | 1 | 1 | 7 | 1 | - | 4 | - |
| > 50 | 14 | - | 3 | - | - | 1 | - | - | 5 | 1 | - | 2 | - | 1 | 1 | - |
| Total | 200 | 12 | 4 | 1 | 8 | 2 | 4 | 20 | 27 | 13 | 18 | 29 | 7 | 24 | 20 | 11 |

N = no thrombus; O = occlusive thrombus; M = mural thrombus

**Table 20. *Lack of relationship between number of severe coronary artery stenoses (≥ 70%) and infarct size (% left ventricular mass) in 200 consecutive acute myocardial infarct cases***

| Infarct Size (%) | Cases | < 69 | ≥ 70 | Lumen Reduction (%) in 1 | 2 | 3 Vessel |
|---|---|---|---|---|---|---|
| ≤ 20 | 97 | 7 | 90 | 39 | 37 | 14 |
| > 20 | 103 | 10 | 93 | 38 | 34 | 21 |
| Total | 200 | 17 | 183 | 77 | 71 | 35 |

P < 0.05 for trend

**Table 21. *Degree and length of maximal stenosis in supplying artery related to presence of acute thrombus in acute myocardial infarcts***

| Luminal Stenosis (%) | Length Stenosis (μm) < 5 | 5-20 | > 20 | Total |
|---|---|---|---|---|
| < 69 | | | | |
| Occlusive thrombus | 1 | 2 | 3 | 6 |
| Mural thrombus | 1 | 3 | 1 | 5 |
| No thrombus | 12 | 6 | 2 | 20 |
| ≥ 70 | | | | |
| Occlusive thrombus | 7 | 26 | 43 | 76 |
| Mural thrombus | 6 | 11 | 14 | 31 |
| No thrombus | 28 | 19 | 15 | 62 |

**Table 22. *Distribution of infarct size (% left ventricular mass) versus survival in 200 consecutive acute infarct cases***

| Survival Days | Cases | Infarct Size (%) < 10 | 11-20 | 21-30 | 31-40 | 41-50 | > 50 |
|---|---|---|---|---|---|---|---|
| < 2 | 70 | 34 | 11 | 11 | 9 | 1 | 4 |
| 3-10 | 74 | 17 | 9 | 24 | 10 | 9 | 5 |
| 11-30 | 56 | 9 | 17 | 9 | 8 | 8 | 5 |
| Total | 200 | 60 | 37 | 44 | 27 | 18 | 14 |

***Table 23. Infarct size (% left ventricular mass), survival time, acute occlusive coronary thrombus***

| Survival Days | ≤ 10 | | 11-20 | | 21-30 | | 31-40 | | 41-50 | | > 50 | | Total | |
|---|---|---|---|---|---|---|---|---|---|---|---|---|---|---|
| | P | T | P | T | P | T | P | T | P | T | P | T | P | T |
| ≤ 2 | 34 | 6 | 11 | 2 | 11 | 4 | 9 | 4 | 1 | 1 | 4 | 3 | 70 | 20 |
| 3-10 | 17 | 2 | 9 | 4 | 24 | 14 | 10 | 6 | 9 | 8 | 5 | 4 | 74 | 38 |
| ≥ 11 | 9 | 4 | 17 | 5 | 9 | 3 | 8 | 2 | 8 | 5 | 5 | 5 | 56 | 24 |
| Total | 60 | 12 | 37 | 11* | 44 | 21 | 27 | 12 | 18 | 14 | 14 | 12 | 200 | 82 |

Infarct Size (%)

P = patients; T = thrombi
* Occlusive thrombus not in supplying artery in one patient.

***Table 24. Infarct size (% left ventricular mass) in relation to its main supplying vessel***

| | | Infarct Size % | | | | | |
|---|---|---|---|---|---|---|---|
| Vessel | Total | ≤ 10 | 11-20 | 21-30 | 31-40 | 41-50 | > 50 |
| Left anterior descending | 105 | 33 | 17 | 20 | 12 | 12 | 11 |
| Left circumflex | 23 | 8 | 5 | 6 | 3 | - | 1 |
| Right coronary artery | 72 | 19 | 15 | 18 | 12 | 6 | 2 |
| Total | 200 | 60 | 37 | 44 | 27 | 18 | 14 |

***Table 25. Heart rupture as a cause of death in 200 consecutive acute myocardial infarcts***

| Source | No Cases | Infarct Size (% Left Ventricular Mass) | | | | | | | Lumen Reduction % | | | | Occlusive Thrombus |
|---|---|---|---|---|---|---|---|---|---|---|---|---|---|
| **AMI 1st** | | < 10 | 11-20 | 21-30 | 31-40 | 41-50 | > 50 | < 69 | ≥ 70 | 1 | 2 | ≥ 3 vessels | |
| + rpt | 31 | 2 | 10 | 11 | 3 | 3 | 2 | 3 | 28 | 15 | 9 | 4 | 15 |
| no | 114 | 30 | 20 | 21 | 21 | 11 | 11 | 13 | 101 | 46 | 40 | 15 | 45 |
| Total | 145 | 32 | 30 | 32 | 24 | 14 | 13 | 16 | 129 | 61 | 49 | 19 | 60 |
| **AMI 2nd** | | | | | | | | | | | | | |
| + rpt | 3 | 1 | - | 2 | - | - | - | - | 3 | 2 | 1 | - | 2 |
| no | 52 | 27 | 7 | 10 | 3 | 4 | 1 | 1 | 51 | 14 | 21 | 16 | 20 |
| Total | 55 | 28 | 7 | 12 | 3 | 4 | 1 | 1 | 54 | 16 | 22 | 16 | 22 |

1st, first episode of IHD; 2nd chronic IHD; rpt, rupture of the heart

***Table 26. Histologic pattern in different types of myocardial necrosis in CHD***

| **Myocardium** | **Coagulation Necrosis (Infarct Necrosis)** | **Coagulative Myocytolysis (Contraction Band or Zenker Necrosis)** | **Colliquative Myocytolysis (Myocytolysis)** |
|---|---|---|---|
| Functional status | Irreversible relaxation ("atonic" death) + stretching by intraventricular pressure | Irreversible contraction ("tetanic" death) | Progressive lost of function ("failing" death) |
| Muscle fiber | Early thinning | Normal or swollen | Increasing edema - vacuolization |
| Nucleus | Elongation-pyknosis progressive karyolysis | Normal | Normal |
| Myofibrils | Elongated sarcomeres in normal registered order, even in late stage | Rhexis - Anomalous irregular cross band formations (coagulation of hypercontracted sarcomeres) | Progressive disappearance "empty cell" (colliquation) |
| Vessels | Secondary wall degeneration and thrombosis | Normal | Normal |
| Infiltration | Massive polymorphonuclear exudation | No early cellular infiltrates. Possible late lymphocytes | No infiltrates |
| Extension-Location | In general unique massive focus of different size. Subendocardial to transmural | Multiple (mono or pluricellular) disseminated or confluent foci of different size in any area | Focal progressively spreading |
| Irreversible in | At least 20-60 min | Few minutes | ? |
| Healing | Removal by Macrophages. Collagenization of Empty Sarcolemmal Tubes | | ? |
| *Frequency in IHD: Acute infarct* | 100% | 100% at outer limit of early infarct<br>85% in myocardium elsewhere | 38% subendo-perivascular |
| Sudden death | 17% histologically demonstrated | 72% (unique demonstrable lesion),<br>86% (including cases with infarct) | 8% |

**Table 27. Type of myocardial necrosis and fibrosis, lumen reduction and acute occlusive thrombus in 208 sudden/unexpected death cases**

| Myocardial Damage | Lumen Reduction % | | | ≥ 70% in | | | Acute Occl. Thr. |
|---|---|---|---|---|---|---|---|
| **Infarct size (%)** | **< 69** | **≥ 70** | **Total** | **1** | **2** | **3 vessels** | |
| ≤ 10 | - | 11 | 11 | 3 | 4 | 4 | 1 |
| 11-30 | - | 16 | 16 | 4 | 8 | 4 | 6 |
| > 30 | 2 | 6 | 8 | 5 | 1 | - | 4 |
| Total | 2 | 33 | 35 | 12 | 13 | 8 | 11 |
| **Coagulative Myocytolysis** | | | | | | | |
| Minimal | 23 | 65 | 88 | 16 | 25 | 24 | 7 |
| Moderate | 6 | 32 | 38 | 8 | 14 | 10 | 9 |
| Extensive | 5 | 18 | 23 | 8 | 9 | 1 | 9 |
| Total | 34 | 115 | 149 | 32 | 48 | 35 | 25 |
| Early | 27 | 68 | 95 | 21 | 25 | 22 | 9 |
| Alveolar | 1 | 27 | 28 | 7 | 9 | 11 | 9 |
| Scarring | 6 | 20 | 26 | 4 | 14 | 2 | 7 |
| Total | 34 | 115 | 149 | 32 | 48 | 35 | 25 |
| **Colliquative Myocytolysis** | | | | | | | |
| Minimal | - | 16 | 16 | 1 | 5 | 10 | 2 |
| **Fibrosis** | | | | | | | |
| No or minimal | 39 | 64 | 103 | 32 | 24 | 8 | 15 |
| Moderate | 4 | 44 | 48 | 13 | 15 | 16 | 4 |
| Extensive | 1 | 56 | 57 | 12 | 15 | 29 | 5 |
| Total | 44 | 164 | 208 | 57 | 54 | 53 | 24 |

**Table 28. Frequency of platelet aggregates and number of occluded intramural arterial vessels in sudden/unexpected death and healthy controls**

| | SD | AD |
|---|---|---|
| **Total cases** | 208 | 97 |
| PA absent | 61 | 23 |
| present in | 147 | 74 |
| 5 | 77 | 35 |
| 5 - 10 | 37 | 14 |
| 11-20 | 22 | 11 |
| 21-30 | 7 | 5 |
| 31-60 vessels | 4 | 9 |
| **Total sections** | 3328 | 1552 |
| PA absent | 2793 | 1273 |
| present in | 535 | 279 |
| 1 | 269 | 124 |
| 2 | 134 | 62 |
| 3 | 61 | 40 |
| 4 | 26 | 21 |
| 5-15 vessels | 45 | 32 |

SD=sudden/unexpected coronary death; AC= accidental death in normal people; PA=platelet aggregates

**Table 29. Frequency of arterial platelet aggregates versus survival time**

| Source Survival (min) | Cases | Platelet Aggregates Absent | Present in | <5 | 5-10 | ≥10 Vessels |
|---|---|---|---|---|---|---|
| SD | | | | | | |
| <10 | 151 | 51 | 100 | 52 | 26 | 22 |
| ≥10 | 57 | 10 | 47 | 25 | 11 | 11 |
| AD | | | | | | |
| <10 | 75 | 22 | 53 | 30 | 11 | 12 |
| ≥10 | 22 | 1 | 21 | 5 | 3 | 13 |

SD=sudden/unexpected coronary death; AD=accidental death in normal people.

**Table 30. Frequency of arterial (AP) and venous (VP) platelet aggregates in relation to intramural blood stasis (total sections), in sudden unexpected death (SD) and healthy control cases (AD)**

| | SD | AD |
|---|---|---|
| No demonstrable stasis | 1005 | 629 |
| + AP | 60 | 48 |
| +VP | 29 | 5 |
| Arterial + venous stasis | 1512 | 620 |
| + AP | 454 | 217 |
| +VP | 418 | 210 |
| Venous stasis alone | 811 | 303 |
| +AP | 21 | 14 |
| +VP | 283 | 107 |
| Total sections | 3328 | 1552 |
| + AP | 535 | 279 |
| + VP | 730 | 322 |

**Table 31. Frequency distribution of medial hyperplasia obliterans (MH) and its localization**

| | | | Localization | | | | | | | Number of Location | | | | |
|---|---|---|---|---|---|---|---|---|---|---|---|---|---|---|
| **Total** | **Cases** | **Cases + MH** | **LV** | **LPM** | **RV** | **RPM** | **IS** | **A** | **CS** | **1** | **2** | **3** | **4** | **5** |
| SD | 208 | 109 | 12 | 73 | 7 | 21 | 46 | 4 | 3 | 66 | 30 | 12 | 1 | - |
| AD | 97 | 76 | 10 | 56 | 5 | 22 | 36 | - | - | 39 | 26 | 8 | 1 | 2 |

LV, RV left and right ventricles; LPM, RPM left and right papillary muscles; IS interventricular septum; A atria; CS conduction system

**Table 32. Medial hyperplasia obliterans (MH) in relation to age**

| Age | SD Cases | MH | AD Cases | + MH |
|---|---|---|---|---|
| < 40 | 35 | 15 | 10 | 7 |
| 40-49 | 33 | 18 | 13 | 9 |
| 50-59 | 68 | 35 | 28 | 24 |
| 60-69 | 45 | 24 | 23 | 18 |
| > 70 | 27 | 17 | 23 | 18 |
| Total | 208 | 109 | 97 | 76 |

SD, sudden/unexpected death; AD, accidental death in normal subjects

*Table 33. Main parameters in cases with congestive heart failure and controls*

| Source | Total Cases | % Men | Age (Years) | Heart Weight (g) | Transverse Heart Diam. (mm) | Ant. LV Wall Thickness (mm) | Coronary Stenosis≥70% | | |
|---|---|---|---|---|---|---|---|---|---|
| | | | | | | | 1 | 2 | ≥3 ves |
| IHD[a] | 63 | 97 | 51±1 | 565±14 | 137±2 | 12±0.5 | 15 | 26 | 16 |
| DC[a] | 63 | 86 | 42±1 | 639±20 | 148±2 | 15±0.5 | 7 | 1 | 1 |
| VP[a] | 18 | 83 | 46±2 | 827±42 | 147±4 | 16±1 | - | 1 | - |
| TH | 46 | 85 | 49±1.5 | 419±19 | 108±11 | 14±1 | 6 | 4 | 1 |
| CHAGAS' | 34 | 76 | 49±2 | 464±28 | 114±3 | 15±0.6 | 1 | - | - |
| AIDS | 38 | 87 | 31±2 | 368±11 | 91±8 | 13±0.6 | 1 | 1 | - |
| BH | 27 | 22 | 58±2 | 427±19 | 107±4 | 18±2 | 4 | 2 | - |
| HT | 45 | 82 | 42±3 | 364±7 | 106±2 | 12±0.4 | - | | |

[a]Hearts excised at transplantation for CHF. AIDS=acquired immunodeficiency syndrome; BH=acute intracranial brain hemorrhage in absence of heart disease; Chagas'=sudden and unexpected death in apparently healthy subjects derum-positive for Chagas' disease; DC=idiopathic dilated cardiomyopathy; HT=accidental head trauma; IHD=ischemic heart disease; TH=transplanted hearts; LV=left ventricle. VP= chronic valvulopathy; SE=standard error. The weight of excised hearts was adjusted adding theroretical atrial weight (see text).

**Table 34. Myocardial damage in CHF cases and controls**

| Source | Cases | Infarct* | Present | CBN** x 100 mm² SE | | Colliquative*** Myocytolysis 0 | 1 | 2 | 3 | MFI | Lymphocytic**** Infiltrates x 100 mm² |
|---|---|---|---|---|---|---|---|---|---|---|---|
| IHD | 63 | 7 | 59 | 1±0.2 | 11±2 | 1 | 14 | 39 | 9 | 17±1 | 0.2±0.05 |
| DC | 63 | - | 54 | 2±1 | 12±4 | 3 | 28 | 28 | 4 | 2±0.3 | 0.4±0.18 |
| VP | 18 | - | 14 | 1±0.5 | 5±2 | 1 | 2 | 14 | 1 | 4±1 | 0.3±0.09 |
| TH | 46 | | 39 | 36±9 | 262±47 | 33 | 11 | 1 | 1 | 1±1 | 6±1 |
| CHAGAS | 34 | - | 17 | 3±1 | 34±16 | 28 | 6 | - | - | 7±1 | 13±2 |
| AIDS | 38 | 4 | 38 | 4±2 | 13±5 | 33 | 3 | 2 | - | 0.04±0.02 | 1.5±0.7 |
| BH | 27 | 1 | 24 | 26±7 | 67±21 | 26 | - | 1 | - | 0.4±0.1 | 2±1.8 |
| HT | 45 | - | 9 | 10±6 | 23±11 | 45 | - | - | - | - | 5±3 |

* Most microfocal subendocardial **CBN, contraction band necrosis or coagulative myocytolysis All stage from early cross bands to healing in all groups; normal head trauma group only early cross bands. ***For grading see method of heart examination; ***only intermyocellular+perivascular infiltrates. IHD, ischemic; DC, dilated cardiomyopathy; AIDS, acquired immunodeficiency syndrome; BH, brain hemorrhage non cardiac patients; HT, normal subjects dying from head trauma; TH, transplanted hearts; MFI, myocardial fibrous index (see text).

***Table 35. Progression of atherosclerotic plaque in relation to increasing intimal thickness and lumen reduction***

| Intimal Thickness μm | Morphologic Variables | Lumen Reduction % |
|---|---|---|
| > 300 | Nodular myosmooth/elastic hyperplasia<br>↓<br>fibrosis | < 50 |
| 600 | proteoglycan deposition in deep intima<br>↓<br>advential/intimal lymph/plasma cells inflammation - media neuritis<br>↓<br>lipoprotein/cholesterol-$Ca^{++}$ salts | 50-69 |
| 1000 | Complications<br>hemorrhage - mural/occlusive thrombosis<br>with or without<br>fissuration/rupture | > 70 |

***Table 36. Old coronary stenosis (≥90% lumen-diameter reduction) without monofocal extensive (≥10% left ventricular mass) myocardial fibrosis (EF)***

| Source | Cases | Stenosis (%) ≥90% | 1 | 2 | ≥3 Vessels |
|---|---|---|---|---|---|
| AMI | 200 | 89 | 68 | 16 | 5 |
| no EF | 145 | 54 | 45 | 6 | 3 |
| SD | 208 | 75 | 51 | 22 | 2 |
| no EF | 133 | 27 | 19 | 7 | 1 |
| NCA | 100 | 40 | 27 | 11 | 2 |
| no EF | 81 | 31 | 21 | 9 | 1 |
| AD | 97 | 6 | 3 | 2 | 1 |
| no EF | 92 | 6 | 3 | 2 | 1 |

AMI, acute myocardial infarct; SD, sudden/unexpected coronary death; NCA, non cardiac atherosclerotic patients; AD, accidental death in normal people.

**Table 37. Platelet aggregates and fibrin-platelet thrombi formed in situ or embolized in sudden coronay death and "controls". Reports in the literature**

| Source | Year | Cases | | Survival | Microvasc. Lesion | Type | No Vessels | Subepi ac. Lesions | Infarct Acute | Infarct Old | Focal Necrosis |
|---|---|---|---|---|---|---|---|---|---|---|---|
| Jorgensen | 1968 | SD | 24 | < 15 m | 4 | PAS | "small number" | 18 | 2 | 0 | ? |
| | | IHD | 78 | < 48 h | 26 | | | 62 | 44 | 48 | ? |
| Haerem | 1972 | S D | 27 | < 10 m | 23 | PAS | 3 (1-22) | 10 | ? | ? | ? |
| | | IHD | 16 | months | 15 | | 2 (1-21) | 5 | ? | ? | ? |
| | | N C | 11 | ? | 6 | | 2 (1-16) | 0 | 0 | 0 | ? |
| Frink | 1978 | S D | 6 | ? | 4 | FPTE | ? | 6 | 0 | 0 | ? |
| | | N C | 3 | ? | 0 | | ? | 0 | 0 | 0 | ? |
| El-Maraghi | 1980 | S D | 50 | < 24 h | 10 | FPTSE | 31±31° | 8 | 4 | 4 | 10 |
| | | IHD | 93 | > 24 h | 9 | | 11±17 | 7 | 9 | 0 | 0 |
| | | Endocar | 5 | > 24 h | 5 | | 36±49 | 0 | 2 | 0 | 6 |
| Falk | 1985 | S D | 25* | < 24 h | 14 | F PTE | 72+ | 25 | 15*** | 0 | 10 |
| Davies | 1986 | S D | 90 | < 6 h | 27 | PAE | ? | 26 | 22** | ? | 23** |
| | | + Unstable angina | 36 | | 16 | | ? | ? | ? | ? | ? |
| | | No unstable angina | 54 | | 11 | | ? | ? | ? | ? | ? |

° mainly in subject <45 yr old; * 26 coronary thrombi in 25 cases; + Frequency of microemboli calculated on 26 perfusion territories: 72 in 29 of 260 histo-sections related to thrombosed arteries and 4 emboli in 4 sections not thrombus-related; ** 23 with focal necrosis PAE in 15 (65%), 22 with infarct necrosis PAE in 16 (75%) 45 without necrosis PAE in 6 (15%); *** not clear it 15 or 20. SD, sudden death; IHD, ischemic heart diseae; NC, non cardiac patients; PA, platelet aggregates; FTP, fibrin-platelet thrombi; PAS, PA formed in situ; PAE, PA embolized; FPT e FPT embolized; FPTS, FPT formed in situ.

**Table 38. Sex and age distribution in sudden coronary death. Reports in the literature**

| Source | Year | | Cases | M | W | M/W | Mean Age (yrs) M | Mean Age (yrs) W |
|---|---|---|---|---|---|---|---|---|
| Bedford | 1933 | | 63 | 57 | 6 | 9 | 61 | 56 |
| Levy | 1936 | | 24 | 21 | 3 | 7 | - | - |
| Munck | 1946 | | 396 | 334 | 62 | 5 | 55 | 63 |
| Rabson | 1948 | | 617 | 581 | 36 | 16 | - | - |
| Croce | 1960 | | 824 | 719 | 105 | 7 | - | - |
| Adelson | 1961 | W | 433 | 389 | 44 | 9 | - | - |
| | | B | 67 | 53 | 14 | 4 | | |
| Franco | 1962 | | 187 | 110 | 77 | 1 | - | - |
| Spiekerman | 1962 | | 94 | 73 | 21 | 3 | - | - |
| Jorgensen | 1968 | | 38 | 26 | 12 | 2 | 60 | 65 |
| Luke | 1968 | W | 59 | 55 | 4 | 14 | | |
| | | B | 19 | 15 | 4 | 4 | | |
| Scott | 1972 | W | 175 | 145 | 30 | 5 | 59 | |
| | | B | 13 | 8 | 5 | 2 | | |
| Titus | 1970 | | 86 | 65 | 21 | 3 | 60 | |
| Friedman | 1973 | | 64 | 58 | 6 | 10 | 53 | |
| Liberthson | 1974 | | 220 | 191 | 29 | 7 | 53 | 60 |
| Lie | 1975 | | 406 | 298 | 108 | 3 | 59 | |
| Haerem | 1975 | | 47 | 33 | 14 | 2 | 62 | 72.5 |
| Margolis | 1975 | | 29 | 27 | 2 | 13 | 54 | |
| Perper | 1975 | W | 109 | 85 | 24 | 3 | - | - |
| | | B | 60 | 48 | 12 | 4 | - | - |
| Reichenbach | 1977 | | 87 | 78 | 9 | 9 | 63 | |
| Baroldi | 1979 | | 208 | 182 | 26 | 7 | 53 | 56 |
| Warnes | 1984 | | 70 | 63 | 7 | 9 | - | - |
| Falk | 1984 | | 25 | 20 | 5 | 4 | - | - |
| Arbustini | 1991 | | 27 | 20 | 7 | 3 | - | - |
| Lie | 1991 | | 202 | 157 | 45 | 3 | - | - |

W, white; B, black; -, not reported

**Table 39. Atherosclerotic lumen reduction and sudden coronary death. Reports in the literature**

| Source | Year | Cases | Stenosis ≥70% (%) | Stenosis ≥70% 1 | 2 | 3 Vess. |
|---|---|---|---|---|---|---|
| Moritz | 1946 | 115 | 84 (73) | - | - | - |
| Adelson | 1961 | 500 | 338 (68) | - | - | - |
| Friedman | 1973 | 59 | 55 (93) | 9 | 11 | 35 |
| Liberthson | 1974 | 220 | 207 (94) | 29 | 54 | 24 |
| Lie | 1975 | 406 | 377 (93) | - | - | - |
| Perper | 1975 | 169 | 153 (91) | 25 | 25 | 103 |
| Reichenbach | 1977 | 87 | 80 (92) | 7 | 16 | 57 |
| Newman | 1982 | 65 | 60 (92) | 11 | 17 | 32 |
| Arbustini | 1991 | 27 | 25 (93) | 3 | 7 | 15 |
| **Total** | | **1648** | **1379 (84)** | **84** | **130** | **366** |
| Baroldi | 1979 | | | | | |
| SD1st | | 133 | 87 (65) | 40 | 34 | 13 |
| SD chr | | 75 | 70 (93) | 13 | 26 | 31 |

SD, sudden/unexpected coronary death; 1st, first episode; chr, chronic CHD

**Table 40. Acute occlusive coronary thrombosis in sudden coronary death. Reports in the literature**

| Source | Year | Cases | | Thrombus Occlusive (%) | | Acute Infarct (%) | AI + Occl. Thrombus (%) |
|---|---|---|---|---|---|---|---|
| Nathanson | 1936 | 142 | | 39 | (27) | 30 (21) | - |
| Lisa | 1939 | 41 | | 10 | (24) | 30 (73) | - |
| Moritz | 1946 | 115 | | 31 | (27) | 15 (13) | - |
| Munck | 1946 | 396 | | 141 | (36) | 74 (19) | 26 (35) |
| Rabson | 1948 | 617 | | 165 | (27) | 31 (5) | - |
| Spain | 1959 | 410 | | 102 | (25) | - | - |
| Branwood | 1956 | 39 | | 14 | (36) | - | - |
| Adelson | 1961 | 500 | | 164 | (33) | 63 (13) | 43 (68) |
| Crawford | 1961 | 75 | | 39 | (52) | 24 (32) | - |
| Franco | 1962 | 187 | | 21 | (11) | 45 (24) | - |
| Jorgensen | 1968 | 24 | | 9 | (38) | 2 (8) | - |
| Friedman | 1973 | 25 | I | 1 | (4) | 0 (0) | - |
| | | 34 | S | 28 | (82) | 7 (21) | - |
| Titus | 1970 | 86 | | 13 | (15) | 34 (39) | - |
| Spain | 1970 | 189 | | 79 | (42) | - | - |
| Roberts | 1979 | 24 | | 2 | (8) | - | - |
| Scott | 1972 | 183 | | 84 | (46) | 118 (64) | 55 (47) |
| Liberthson | 1975 | 220 | | 70 | (32) | 59 (27) | - |
| Lie | 1975 | 406 | | 69 | (17) | 148 (36) | - |
| Kuller | 1975 | 118 | | 30 | (25) | 13 (11) | - |
| Baba | 1975 | 121 | | 64 | (53) | 50 (41) | 11 (22) |
| Reichenbach | 1977 | 87 | | 9 | (10) | 22 (15) | - |
| Newman | 1982 | 65 | | 21 | (32) | 6 (9) | - |
| Warnes | 1984 | 70 | | 11 | (16) | 0 (0) | - |
| Davies | 1984 | 100 | | 44 | (44) | - | - |
| Kragel | 1991 | 21 | | 6 | (29) | - | - |
| Arbustini | 1991 | 27 | | 7 | (26) | - | - |
| Lie | 1991 | 202 | | 36 | (18) | 63 (31) | - |
| **Total** | | **4524** | | **1309** | **(29)** | **834 (22) *** | **-** |
| Baroldi | 1979 | | | | | | |
| SD 1st | | 133 | | 11 | (8.) | 16 (12) | 8 (50) |
| SD chr | | 75 | | 21 | (28) | 19 (25) | 3 (16) |

I instantaneous sudden death; S sudden death within 24 hours - AI Acute infarct
* Percentage calculated on the total of 3714 cases in whom the frequency of an acute infarct has been reported.

***Table 41. Frequency of pathological heart weight in sudden coronary death. Reports in the literature***

| Source | Year | Cases | Heart Weight (500 g) | |
|---|---|---|---|---|
| | | | < | ≥ |
| Levy | 1936 | 24 | 4 | 20 |
| Nathanson | 1936 | 139 | 104 | 35 |
| Moritz | 1946 | 115* | 115 | 0 |
| Adelson | 1961 | 500 | 170 | 330 |
| Crawford | 1961 | 75 | 65 | 10 |
| Titus | 1970 | 86 | 41 | 45 |
| Scott | 1972 | 183 | 93 | 90 |
| Kuller | 1975 | 118 | 76 | 42 |
| Haerem | 1975 | 33 | 15 | 18 |
| Frink | 1978 | 6 | 6 | 0 |
| **Total** | | **1279** | **689** | **590** |
| Baroldi | 1979 | | | |
| SD 1st | | 133 | 76 | 57 |
| SD chr | | 75 | 18 | 57 |

* All soldiers less than 40 yrs old. For legends see Table 39.

***Table 42. Intimal hemorrhage in sudden coronary death. Reports in the literature***

| Source | Year | Total Cases | | Hemorrhage Alone | + Rupt Plaque | + Thrombus | Rupt-Th |
|---|---|---|---|---|---|---|---|
| Jorgensen | 1968 | 24 | | 1 | ? | ? | ? |
| Friedman | 1973 | 25 | I | 0 | 0 | 0 | 0 |
| | | 34 | S | 0 | 0 | 0 | 19 |
| Liberthson | 1974 | 220 | | 22 | ? | ? | ? |
| Kuller | 1975 | 169 | | 9 | ? | ? | ? |
| Baba | 1975 | 121 | | 13 | 27 | 9 | 0 |
| Kragel | 1991 | 21 | | 0 | 0 | 0 | 4 |
| Arbustini | 1991 | 27 | | 0 | 0 | 0 | 0 |
| **Total** | | **641** | | **45** | **27** | **9** | **23** |

***Table 43. Collective pathology from reports in the literature in 78 athletes who died suddenly***

| Pathology | M | W | Total | Age | Heart Weight | History | Physical Activity | Other |
|---|---|---|---|---|---|---|---|---|
| Anomalous origin cor. artery | 7 | - | 7 | 17-22 | 450-480 | no | 6 | 1 + hypertr. cardiomyopathy |
| | | | | 19 | 465 | | | |
| Hypoplasia right cor. artery | - | 1 | 1 | 17 | 280 | no | no | - |
| "Mural" left ant. descending branch | - | 1 | 1 | 17 | 260 | no | 1 | Reflow necrosis coag. myocyt. |
| Hypertrophic cardiomyopathy | 13 | 2 | 15 | 13-30 | 360-650 | 7 | ? | Thickening septal arteries |
| | | | | 18 | 561 | | | |
| Floppy mitral valve | 3 | - | 3 | 17-27 | 420-530 | 2 | 2 | 1+fibromusc.hyper. |
| | | | | 22 | 475 | | | AV node artery |
| Idiopathic LV hypertrophy | 4 | - | 4 | 16-28 | 420-465 | no | 4 | 1+fibromusc.hyper. |
| | | | | 19 | 448 | | | AV node artery |
| Hearttumor | 1 | 1 | 2 | 12-15 | ? | 1 | 2 | one fibroma |
| | | | | 14 | | | | one mixoma |
| Rupture aorta | 2 | - | 2 | 18-21 | ? | 2 | 2 | Medial cystic necrosis |
| | | | | 19 | | | | |
| Lung thrombo-embolism | - | 1 | 1 | 20 | ? | no | ? | - |
| Anomaly conduct.system | 1 | 1 | 2 | 11-35 | ? | no | ? | - |
| | | | | 23 | | | | |
| Right ventr. dysplasia | 7 | - | 7 | 16-26 | 410-540 | 4 | 7 | - |
| | | | | 21 | 502 | | | |
| Atherosclerosis coronary | 33 | - | 33 | 17-58 | 345-480 | 16 | 14 | - |
| | | | | 38 | 406 | | | |

Data collected from Heath, 1969, Opie, 1975; Noakes et al 1979; Maron et al 1980; Morales et al, 1980; Waller et al, 1980, Tsung et al 1982; Virmani et al, 1982; Voigt et al, 1982; Thiene et al, 1985.

***Table 44. Collective cardiovascular pathology from reports in the literature in 78 athletes who died suddenly related to their age***

| Pathology | < 20 | 20-29 | 30-39 | > 40 | Total |
|---|---|---|---|---|---|
| Anomalous origin cor. arteries | 5 (17%) | 2 (8%) | - | - | 7 (9%) |
| Hypoplasia right cor. artery | 1 (3) | - | - | - | 1 (1) |
| "Mural" left ant. desc. branch | 1 (3) | - | - | - | 1 (1) |
| Hypertrophic cardiomyopathy | 10 (34) | 4 (16) | 1 (17) | - | 15 (19) |
| Floppy mitral valve | 1 (3) | 2 (8) | - | - | 3 (4) |
| Idiopathic LV hypertrophy | 3 (10) | 1 (4) | - | - | 4 (5) |
| Heart tumor | 2 (7) | - | - | - | 2 (3) |
| Rupture aorta | 1 (3) | 1 (4) | - | - | 2 (3) |
| Long thrombo-embolism | - | 1 (4) | - | - | 1 (1) |
| Anomaly conduct. system | 1 (3) | - | 1 (17) | - | 2 (3) |
| Right ventr. dysplasia | 3 (10) | 4 (16) | - | - | 7 (9) |
| Atherosclerosis coronary | 1 (3) | 10 (40) | 4 (66) | 18 (100) | 33 (42) |
| **Total** | **29 (100)** | **25 (100)** | **6 (100)** | **18 (100)** | **78 (100)** |

References quoted in the text and table 45.

***Table 45. Collective pathology from reports in the literature in 36 joggers who died suddenly***

| Pathology | M | W | Total | Age (yrs) | Heart Weight (g) | History | Physical Activity |
|---|---|---|---|---|---|---|---|
| G.I. hemorrhage | 1 | - | 1 | 28 | ? | no | ? |
| "Mural" coronary artery | 2 | - | 2 | 34-54<br>44 | 400-460<br>430 | 1 | 2 |
| Atherosclerosis coronary | 33 | - | 33 | 18-58<br>42 | ? | 13 | 33 |

Data from Morales et al, 1982; Virmani et al, 1982; Thiene et al, 1985.

**Table 46. Silent acute myocardial infarct in sudden/unexpected death (SD) versus clinical infarct cases and SD without histologic infarct**

| Morphologic Variables | Silent Infarct 35 Cases | Clinical Infarct 200 Cases | SD no Infarct 173 Cases |
|---|---|---|---|
| ***Coronary stenosis*** | | | |
| < 69 | 3 | 17 | 48 |
| ≥70 | 32 | 183 | 125 |
| ≥70 in | | | |
| 1 | 12 | 77 | 41 |
| 2 | 12 | 71 | 48 |
| 3 or more vessels | 8 | 35 | 36 |
| ***Occlusive thrombus*** | | | |
| acute | 11 | 83 | 21 |
| organized | 2 | 30 | 12 |
| ***Acute infarct*** | | | |
| size (%) | | | |
| < 20 | 20 | 97 | - |
| ≥ 20 | 8 | 103 | - |
| age (days) | | | |
| < 2 | 19 | 70 | - |
| 2-10 | 4 | 74 | - |
| 11-30 | 12 | 56 | - |
| location | | | |
| subendocardial | 6 | 26 | - |
| internal half | 18 | 52 | - |
| transmural | 11 | 122 | - |
| ***Myocardial fibrosis*** | | | |
| absent/microfocal | 31 | 145 | 148 |
| extensive | 4 | 55 | 25 |
| ***Heart weight*** | | | |
| < 500 g | 23 | 96 | 98 |
| ≥ 500 | 12 | 104 | 75 |

***Table 47. Frequency and extent of contraction band necrosis in different groups***

| **Source** | **Contraction Band Necrosis** | | | | | | |
|---|---|---|---|---|---|---|---|
| | **Cases** | **Present** | **Foci**** | **Myocells**** | **Histologic Stages of CBN** | | |
| | | | | | **Cross Band** | **Alveolar** | **Healing** |
| Coronary | 25 | 19 | 28±8 | 508±200 | 3 | 13 | 3 |
| *+ infarct silent** | *4* | *4* | *29±10* | *1717±698* | *1* | *2* | *1* |
| *no infarct* | *21* | *15* | *27±10* | *185±48* | *2* | *11* | *2* |
| *no resuscitation* | *10* | *6* | *9±2* | *102±59* | *1* | *5* | - |
| *resuscitation* | *11* | *9* | *39±16* | *241±65* | *1* | *6* | *2* |
| Chagas | 34 | 34 | 3±1 | 34±16 | 8 | 16 | 10 |
| BH | 27 | 24 | 26±7 | 67±21 | 13 | 9 | 2 |
| *≤ 1 day survival* | *14* | *12* | *16±5* | *26±29* | *9* | *3* | - |
| *>1 day survival* | *13* | *12* | *37±14* | *108±134* | *4* | *6* | *2* |
| TRH | 46 | 39 | 36±9 | 262±47 | 14 | 17 | 8 |
| AIDS | 38 | 25 | 4±2 | 13±5 | 19 | 6 | - |
| CHF | 144 | 126 | 2±0.3 | 11±2 | 65 | 25 | 36 |
| Cocaine | 26 | 11 | 4±1 | 11±4 | 11 | - | - |
| Head trauma | 45 | 9 | 1±6 | 23±11 | 9 | - | - |
| ≤ 1 hour survival | 26 | 1 | 0.5 | 35 | 1 | - | - |
| *>1 hour survival* | *19* | *8* | *12±6* | *21±12* | *8* | - | - |
| Electrocution | 21 | 1 | 8 | 46 | 1 | - | - |
| Co intoxication | 26 | 3 | 1±0.5 | 5±2 | 3 | - | - |

* The four cases with silent acute myocardial infarction excluded when CBN extent was calculated in coronary group. **Number of foci and myocells + CBN, standard error.

| **Source/** | | **Contraction Band Necrosis** | | | | | | | |
|---|---|---|---|---|---|---|---|---|---|
| **No. of Cases** | | **Present** | **LV** | **LV+RV** | **LV+IVS** | **LV+RV+IVS** | **RV** | **RV+IVS** | **IVS** |
| Coronary | 25 | 19 | 1 | 2 | 4 | 11 | 1 | - | - |
| Chagas | 34 | 34 | 6 | 10 | 2 | 14 | - | - | 2 |
| BH | 27 | 24 | 3 | 1 | 6 | 14 | - | - | - |
| TRH | 46 | 39 | 2 | 1 | 2 | 31 | 1 | 1 | 1 |
| AIDS | 38 | 25 | 6 | 2 | 2 | 8 | 5 | - | 2 |
| CHF | 144 | 126 | 20 | 18 | 9 | 67 | 5 | 4 | 3 |

LV, left ventricle free wall; RV, right ventricle free wall; IVS, interventricular septum.
* Cocaine and AD groups excluded because sampling limited to anterior LV (see methods).

**Table 48. Colliquative myocytolysis in relation to extensive (≥ 10%) myocardial fibrosis (F) in different conditions**

| Source | Cases | Colliquative Myocytolysis (Grades) 0 | (F) | 1 | (F) | 2 | (F) | 3 | (F) |
|---|---|---|---|---|---|---|---|---|---|
| Congestive heart failure * | 144 | 5 | (3) | 44 | (14) | 81 | (37) | 14 | (9) |
| *Coronary heart disease* | *63* | *1* | *(1)* | *14* | *(11)* | *39* | *(34)* | *9* | *(8)* |
| *Dilated cardiomyopathy* | *63* | *3* | *(1)* | *28* | *(3)* | *28* | *(-)* | *4* | *(-)* |
| *Valvular disease* | *18* | *1* | *(1)* | *2* | *(-)* | *14* | *(3)* | *1* | *(1)* |
| Sudden/unexpected death | 59 | 47 | (13) | 12 | (8) | - | (-) | - | |
| *Coronary* | *25* | *19* | *(5)* | *6* | *(5)* | - | *(-)* | - | |
| *Silent Chagas' disease* | *34* | *28* | *(8)* | *6* | *(3)* | - | *(-)* | - | |
| Transplanted hearts | 46 | 33 | (-) | 11 | (3) | 1 | (-) | 1 | (-) |
| *Survival (days) < 7* | *9* | *8* | *(-)* | *1* | *(-)* | - | *(-)* | - | |
| *7-30* | *10* | *10* | *(-)* | - | *(-)* | - | *(-)* | - | |
| *31-365* | *13* | *9* | *(-)* | *4* | *(1)* | - | *(-)* | - | |
| *> 365* | *14* | *6* | *(-)* | *6* | *(2)* | *1* | *(-)* | *1* | *(-)* |
| Brain hemorrhage | 27 | 26 | (-) | - | (-) | 1 | (-) | - | |
| AIDS | 38 | 33 | (-) | 3 | (-) | 2 | (-) | - | |
| Cocaine abusers | 26 | 26 | (3) | - | (-) | - | (-) | - | |
| Carbon monoxide intoxication | 26 | 26 | (-) | - | (-) | - | (-) | - | |
| Head trauma | 45 | 45 | (1) | - | (-) | - | (-) | - | |
| Electrocution | 21 | 21 | (-) | - | (-) | - | (-) | - | |

* Hearts excised at transplantation; (F) + extensive fibrosis

***Table 49. Frequency and extent of "pathological" (≥ 20% x histological area) myocardial disarray in different conditions***

| Source | Cases | Myocardial Disarray Sites + | | | | | | | | |
|---|---|---|---|---|---|---|---|---|---|---|
| | | **0** | **1** | **2** | **3** | **4** | **5** | **6** | **7** | **8** |
| Intracranial brain hemorrhage | 27 | 12 | - | 4 | 1 | 3 | 1 | 1 | - | 5 |
| Sudden/unexpected coronary death | 25 | 13 | 2 | - | 2 | 1 | 2 | - | 2 | 3 |
| Transplanted hearts | 46 | 25 | 3 | 1 | 2 | 4 | 3 | 0 | 4 | 4 |
| *survival < 7* | *9* | *9* | - | - | - | - | - | - | - | - |
| *7-30* | *10* | *4* | *1* | *1* | - | *3* | - | - | *1* | - |
| *31-365* | *13* | *6* | - | - | *1* | - | *3* | - | *2* | *1* |
| *> 365* | *14* | *6* | *2* | - | *1* | *1* | - | - | *1* | *3* |
| Sudden/unexp. death silent Chagas | 34 | 25 | 4 | 4 | - | 1 | - | - | - | - |
| Cocaine abusers ** | 26 | 22 | 4 | - | - | - | - | - | - | - |
| Congestive heart failure * | 144 | 124 | 10 | 5 | 2 | - | 3 | - | - | - |
| AIDS | 38 | 34 | - | - | 2 | 1 | - | - | - | 1 |
| Carbon monoxide intoxication ** | 26 | 26 | - | | | | | | | |
| Head trauma ** | 45 | 45 | - | | | | | | | |
| Electrocution ** | 21 | 21 | - | | | | | | | |

* Hearts excised at transplantation ** Only anterior left ventricle available for study.
+ Site corresponds to myocardial samples in each heart: anterior, lateral, posterior of left and right ventricle, anterior and postero interventricular septum.

***Table 50. Frequency of endocardial myoelastofibrosis in different conditions***

| Source | Cases | Absent | Endocardial Myofibroelastosis: Nodular Myo Hyper-Plasia | Myofibro-elastosis | LV | RV | IV |
|---|---|---|---|---|---|---|---|
| Sudden/unexpected coronary death | 25 | 13 | 2 | 10 | 6 | 6 | 9 |
| Transplanted hearts | 46 | 29 | 6 | 11 | 11 | 5 | 6 |
| Congestive heart failure * | 144 | 3 | 15 | 126 | 129 | 29 | 125 |
| Sudden/unexpected death silent Chagas' | 34 | 15 | 1 | 18 | 17 | 6 | 10 |
| Intracranial brain hemorrhage | 27 | 10 | 3 | 14 | 13 | 4 | 9 |
| AIDS | 38 | 16 | 7 | 15 | 11 | 5 | 11 |
| Cocaine abusers ** | 26 | 19 | 5 | 2 | 2 | - | - |
| Carbon monoxide intoxication ** | 26 | 26 | - | - | - | - | - |
| Head trauma ** | 45 | 45 | - | - | - | - | - |
| Electrocution ** | 21 | 21 | - | - | - | - | - |

*Hearts excised at transplantation ** Only anterior left ventricle available for study; LV, left ventricle; RV right ventricle; IV, interventricular septum

# References

1. Ablad B. Beta adrenergic mechanisms and atherogenesis. In Hypertension -The tip of an iceberg. Manchester: Adis Press, 1988:23.
2. Adam RD. Diseases of muscle. A study in pathology. London: Harper & Row, 1975:452.
3. Adelson L, Hoffman W. Sudden death from coronary disease related to a lethal mechanism arising independently of vascular occlusion or myocardial damage. JAMA 1961; 176:129.
4. Ahrens PJ, Sheehan FH, Dahl J et al. Extension of hypokinesia into angiographically perfused myocardium in patients with acute infarction. J Am Coll Cardiol 1993; 22:1010.
5. Airaksinen KEJ, Tahvanain KUD, Ecberg DE et al. Arterial baroreflex impairment in patients during acute coronary occlusion. J Am Coll Cardiol 1998; 32:1641.
6. Aldridge HE, Trimble AS. Progression of proximal coronary artery lesions to total occlusion after aorta-coronary saphenous vein by pass grafting. J Thorac Cardiov Surg 1971; 62:7.
7. Algra A, Tijssen JGP, Roelandt JRTC et al. Heart rate variability from 24-hour electrocardiography and the 2-year risk for sudden death. Circulation 1993; 88:180.
8. Alonzo DR, Starek PK, Minick CR. Induction of atherosclerosis in transplanted rabbit hearts by the sinergy of graft rejection and cholesterol-rich diet. Circulation 1970; 42 (Suppl 3):5.
9. Ambrose JA, Tannenbaum MA, Alexopoulos D et al. Angiographic progression of coronary artery disease and the development of myocardial infarction. J Am Coll Cardiol 1988; 12:56.
10. Ambrose JA, Winters SL, Arora RR et al. Coronary angiographic morphology in myocardial infarction: a link between the pathogenesis of unstable angina and myocardial infarction. J Am Coll Cardiol 1985; 6:1233.
11. Ambrose JA, Winters SL, Stern A et al. Angiographic morphology and pathogenesis of unstable angina pectoris. J Am Coll Cardiol 1985; 5:609.
12. Ambrose JA. Plaque disruption and the acute coronary syndromes of unstable angina and myocardial infarction: if the substrate is similar why is the clinical presentation different? J Am Coll Cardiol 1992; 19:1653.
13. Amsterdam EA, Pan H, Rendig SV et al. Limitation of myocardial infarct size in pigs with a dual lipoxygenase-cyclooxygenase blocking agent by inhibition of neutrophil activity without reduction of neutrophil migration. J Am Coll Cardiol 1993; 22:1738.
14. Anderson JL, Rodier HE, Green LS. Comparative effects of beta-adrenergic blocking drugs on experimental ventricular fibrillation threshold. Am J Cardiol 1983; 51:1196.
15. Anderson KP, Stinson EB, Derby GC et al. Vulnerability of patients with obstructive hypertrophic cardiomyopathy to ventricular arrhythmia induction in the operating room. Am J Cardiol 1983; 51:812.
16. Angelini A, Benciolini P, Thiene G. Radiation-induced coronary obstructive atherosclerosis and sudden death in a teenager. Inter J Cardiol 1985; 9:371.
17. Angelini A, Thiene G, Frescura G et al. Coronary arterial wall and atherosclerosis in youth (1-20 years): a histologic study in northern Italian population. Intern J Cardiol 1990; 28:361.
18. Angerer P, Siebert U, Kothny W et al. Impact of social support, clinical hostility and anger expression on progression of coronary atherosclerosis. J Am Coll Cardiol 2000; 36:1781.
19. Antoci B, Baroldi G, Costanzi G et al. Atherosclerotic plaque in heart and brain arteries. In: Baroldi G, Bonomo L, Fieschi C, Maseri A, eds. Atherosclerosis-1st international workshop on ischemic and brain disease. Rome: Pensiero Scientifico, 1980:89.
20. Araie E, Fujita M, Ejiri M et al. Relation between the preexistent coronary collateral circulation and recanalization rate of intracoronary thrombolysis for acute myocardial infarction. J Am Coll Cardiol 1990; 15:218A.
21. Arani DT, Greene DG, Bunnel IL et al. Reductions in coronary flow under resting conditions in collateral dependent myocardium of patients with complete occlusion of the left anterior descending coronary artery. J Am Coll Cardiol 1984; 3:668.
22. Arbustini E, Grasso M, Diegoli M et al. Coronary atherosclerotic plaques with and without thrombus in ischemic heart syndromes: a morphologic, immunoistochemical and biochemical study. Am J Cardiol 1991; 68:36B.
23. Armour AJ. Myocardial ischaemia and the cardiac nervous system. Cardiov Res 1999; 41:41-54.
24. Ashraf M, White F, Bloor CM. Ultrastructural influence of reperfusing dog myocardium with calcium-free blood after coronary artery occlusion. Am J Pathol 1978; 90:423.

25. Baba N, Bashe WJ, Keller MD et al. Pathology of atherosclerotic heart disease in sudden death. Organizing thrombosis and acute coronary vessel lesions. Circulation 1975; 53(Suppl 3); 51.
26. Bakker A, Thone F, Jacob W. Contact sites between the inner and outer mitochondrial membrane in standard versus hibernating myocardium. Cardiovasc Pathol 1995; 4:195.
27. Barber MJ, Mueller TM, Henry DP et al. Transmural myocardial infarction in the dog produces sympathectomy in noninfarcted myocardium. Circulation 1983; 67:787.
28. Bardales RH, Hailey SL, Su Su Xie et al. In situ apoptosis assay for detection of early acute infarction. Am J Pathol 1996; 149:821.
29. Barger AC, Beeuwkes R, Lainey LL et al. Hypothesis: Vasa vasorum and neovascularization of human coronary arteries. A possible role in the pathophysiology of atherosclerosis. N Engl J Med 1984; 310(3):175.
30. Barilli F, De Vincentis G, Maugieri E et al. Recovery of contractility of viable myocardium during inotropic stimulation is not dependent on an increase of myocardial blood flow in the absence of collateral filling. J Am Coll Cardiol 1999; 33:697.
31. Barnes AR, Ball RG. The incidence and situation of myocardial infarction in one thousand consecutive postmortem examination. Am J Med Sc 1932; 183:215.
32. Baroldi G, Corallo S, Maroni M et al. Focal lymphocytic myocarditis in acquired immunodeficiency syndrome (AIDS): a correlative morphologic and clinical study in 26 consecutive fatal cases. J Am Coll Cardiol 1988; 12:463.
33. Baroldi G, Di Pasquale G, Silver MD et al. Type and extent of myocardial injury related to brain damage and its significance in heart transplantation: a morphometric study. J Heart Lung Transplant 1997; 16:994.
34. Baroldi G, Falzi G, Lampertico P. The nuclear patterns of cardiac muscle fiber. Cardiologia 1967; 51:109.
35. Baroldi G, Falzi G, Mariani F et al. Morphology, frequency and significance of intramural arterial lesions in sudden coronary death. G Ital Cardiol 1980; 10:644.
36. Baroldi G, Falzi G, Mariani F. Sudden coronary death. A postmortem study in 208 selected cases compared to 97 "control" subjects. Am Heart J 1979; 98:20.
37. Baroldi G, Fineschi V, Silver MD et al. Myocardial contraction bands. Definition, quantification and significance in human pathology. Int J Legal Med 2001; 14-30.
38. Baroldi G, Silver MD, Fineschi V et al. Frequency and extent of contraction band necrosis in orthotopically transplanted human hearts. Int J Cardiol 2003; 88:676.
39. Baroldi G, Manion WC. Microcirculatory disturbances and human myocardial infarction. Am Heart J 1967; 74:171.
40. Baroldi G, Mantero O, Scomazzoni G. The collaterals of the coronary arteries in normal and pathologic hearts. Circ Res 1956; 4:223.
41. Baroldi G, Marzilli M, L'Abbate A et al. Coronary occlusion: cause or consequence of acute myocardial infarction? A case report. Clin Cardiol 1990; 13,49.
42. Baroldi G, Milam JD, Wakasch DC et al. Myocardial cell damage in "stone heart". J Mol Cell Cardiol 1974; 6:395.
43. Baroldi G, Oliveira SJM, Silver MD. Sudden and unexpected death in clinically "silent" Chagas'disease. A hypothesis. Int J Cardiol 1997; 58:263-68.
44. Baroldi G, Parravicini C, Gaiera G. Sudden cardiac death in a "silent" case of acquired immunodeficiency syndrome (AIDS). G Ital Cardiol 1993; 23:353.
45. Baroldi G, Radice F, Schmidt C et al. Morphology of acute myocardial infarction in relation to coronary thrombosis. Am Heart J 1974; 87:65.
46. Baroldi G, Rapezzi C, De Maria R et al. The clinicopathologic spectrum of hypertrophic cardiomyopathy. The experience of the Italian Heart Transplant Program. In: Camerini F, Gavazzi A, De Maria R, eds. Advances in Cardiomyopathies. Heidelberg: Springer 1998.
47. Baroldi G, Scomazzoni G. Coronary Circulation in the Normal and Pathologic Heart. American Registry of Pathology, A.F.I.P. ed. Washington D.C.: U.S. Government Printing Office, 1967.
48. Baroldi G, Silver MD, De Maria R et al. Lipomatous metaplasia in left ventricular scar. Can J Cardiol 1997; 13:65.

49. Baroldi G, Silver MD, De Maria R et al. Pathology and pathogenesis of congestive heart failure. A quantitative morphologic study of 144 hearts excised at transplantation. Pathogenesis 1998; 1:107.
50. Baroldi G, Silver MD, Lixfield W et al. Irreversible myocardial damage resembling catecholamine necrosis secondary to acute coronary occlusion in dogs: its prevention by propranolol. J Molec Cell Cardiol 1977; 9:687.
51. Baroldi G, Silver MD, Mariani F et al. Correlation of morphological variables in the coronary atherosclerotic plaque with clinical patterns of ischemic heart disease. Am J Cardiov Path 1988; 2:159.
52. Baroldi G, Silver MD. The healing of myocardial infarcts in man. Giorn It Cardiol 1975; 5:465.
53. Baroldi G. Medial hyperplasia obliterans of the intramural coronary arterial vessels. G Ital Cardiol 1986; 16:537.
54. Baroldi G. Myocardial cell death, including ischemic heart disease and its complications. In Silver MD, Gotlieb A, Schoen FJ, eds. Cardiovascular Pathology, 3rd Edition, New York: Saunders WH Co 2001:504.
55. Baroldi G. Structural background and clinical imaging: is pathology still needed? In: Marzullo P, ed. "Understanding cardiac imaging techniques from basal pathology to image fusion". Nato Advanced Research Workshop, IOS Press, Amsterdam, 2001.
56. Baroldi G. The cardiovascular pathologist: a pathologist awaiting a better definition. Cardiov Pathol 1993; 2,215.
57. Baroldi G. Acute coronary occlusion as a cause of myocardial infarct and sudden coronary heart death. Am J Cardiol 1965; 16:859.
58. Baroldi G. Coronary stenosis: ischemic or non ischemic factor? Am Heart J 1978; 96:139.
59. Baroldi G. Coronary thrombosis: facts and beliefs. Am Heart J 1976; 91:683.
60. Baroldi G. Different types of myocardial necrosis in coronary heart disease: a pathophysiological review of their functional significance. Am Heart J 1975; 89:742.
61. Baroldi G. Diseases of the extramural coronary arteries. In "Cardiovascular Pathology", MD Silver ed. Churchill Livingstone Inc., New York, 2nd ed. 1991, p. 487.
62. Baroldi G. High resistance of the human myocardium to shock and red blood cells aggregation (sludge). Cardiologia 1969; 54:271.
63. Baroldi G. Limitation of infarct size: Theoretical aspects. In: Cardiology. An International Perspective. Chazov EI, Smirnov VN, Organov RG, eds. Vol. 1. New York: Plenum Press, 1984:517.
64. Baroldi G. Morpho-pathological contribution to the modern interpretation of ischemic heart disease (V Intern Symposium of Atherosclerosis, Bologna, Italy 1983). Lenzi S, Deschovich G, eds. Lancaster: MTP Press, 1985:279.
65. Baroldi G. Morphologic forms of myocardial necrosis related to myocardial cell function. In: Silver MD, ed. Cardiovascular Pathology. 2nd ed. New York: Churchill Livingstone Pub., 1991:643.
66. Baroldi G. Pathology and mechanisms of sudden death. In: Hurst JW, ed. The Heart. 5th and 6th ed. New York: McGraw-Hill Book Co., 1982,1985:529.
67. Baroldi G. The coronary arteries in cor pulmonale. Acta Cardiol 1971; 26:602.
68. Baroldi G. The coronary circulation in man. In: CJJ Schwartz, NT Wherthesse, S Wolf, eds. Structure and Function of the Circulation. Vol. 2 New York: Plenum Publi. Corp., 1981:77.
69. Barrie HJ, Urback PG. The cellular changes in myocardial infarction. Canad Med Ass J 1957; 77:106.
70. Barry JM, Roberts WC, McAllister HA et al. Sudden death in young athletes. Circulation 1980; 62:2.
71. Bashe WJ, Baba N, Keller MD et al. Pathology of atherosclerotic heart disease in sudden death. II The significance of myocardial infarction. Circulation 1975; 63:51(Suppl 3).
72. Basso C, Thiene G, Corrado D et al. Hypertrophic cardiomyopathy and sudden death in the young: pathological evidence of myocardial ischemia. Hum Pathol 2000; 31:988.
73. Basuk WL, Reimer KA, Jennings RB. Effect of repetitive brief episodes of ischemia on cell volume, electrolytes and ultrastructure. J Am Coll Cardiol 1986; 8:33A.
74. Batsakis JG. Degenerative lesions of the heart. In: Gould SE, ed. The Pathology of the Heart and Blood Vessels. 3rd ed. Springfield: Charles C Thomas Pub., 1968:479.
75. Bayés de Luna A, Coumel P, Leclercq JF. Ambulatory sudden cardiac death: mechanisms of production of fatal arrhythmia on the basis of data from 157 cases. Am Heart J 1989; 117:151.

76. Bayés de Luna A, Guindo J. Sudden Cardiac Death. Editorial MCR, Barcelona,1989.
77. Bayés de Luna A, J Guindo, Borja M et al. Recasting the approach to the treatment of potentially malignant ventricular arrhythmias after the cast study. Cardiov Drugs Ther 1990; 4:651.
78. Bayés de Luna A, J Guindo. Sudden death in ischemic heart disease Rev Port Cardiol 1990; 9(5):473.
79. Bean SL, Saffitz JE. Transmural heterogeneity of norepinephrine uptake in failing human hearts. J Am Coll Cardiol 1994; 23:579-85.
80. Bean WB. Bullet wound of the heart with coronary artery ligation. Am Heart J 1944; 21:375.
81. Bedford THB. The pathology of sudden death: a review of 198 cases "brought in dead". J Path Bact 1933; 36:333.
82. Beltrami AP, Urbanek K, Kajstura J et al. Evidence that human cardiac myocytes divide after myocardial infarction. N Engl J Med 2001; 344:1750.
83. Beltrami CA, Finato N, Rocco M et al. Structural basis of end stage failure in ischemic cardiomyopathy in humans. Circulation 1994; 89:151.
84. Bemis CE, Gorlin R, Kemp HG et al. Progression of coronary artery disease. A clinical arteriographic study. Circulation 1973; 47:455.
85. Benditt EP. Evidence for a monoclonal origin of human atherosclerotic plaques and some implications. Circulation 1974; 50:650.
86. Bengel FM, Neberfuhr P, Ziegler SI et al. Serial assessment of sympathetic reinnervation after orthotopic heart transplantation. A longitudinal study using PET and C-11 hydroxyephedrine. Circulation 1999; 99:1866.
87. Benshop RJ, Nieuwhenhuis EES, Tromp BA et al. Effects of b-adrenergic blockade on immunologic and cardiovascular changes induced by mental stress. Circulation 1994; 89:762.
88. Benson DW, Benditt DG, Anderson RW et al. Cardiac arrest in young, ostensibly healthy patients: clinical, hemodynamic, and electrophysiologic findings. Am J Cardiol 1983; 52:65.
89. Berry W. The unsettling of America culture and agriculture. San Francisco: Sierra Club Books, 1977.
90. Bertrand ME, LaBlanche JM, Tilmant PY et al. Frequency of provoked coronary arterial spasm in 1089 consecutive patients undergoing coronary arteriography. Circulation 1982; 65:1299.
91. Bevilacqua M, Norbiato G, VagoT et al. Alterations in norepinephrine content and beta adrenoceptor regulation in myocardium bordering aneurysm in human heart: their possible role in the genesis of ventricular tachycardia. Eur J Clin Invest 1986; 16:163.
92. Bieber CP, Stinson EB, Shumway NE et al. Cardiac transplantation in man. VII Pathology. Am J Cardiol 1970; 25:84.
93. Billinger M, Fleish M, Eberli FR et al. Is the development of myocardial tolerance to repeated ischemia in humans due to preconditioning or to collateral recruitment? J Am Coll Cardiol 1999; 33:1027.
94. Billingham ME. The postsurgical heart. The pathology of cardiac transplantation. Am J Cardiov Path 1988; 1:319.
95. Bing RJ, Tillmanns H, Fauvel JM et al. Effect of prolonged alcohol administration on calcium transport in heart muscle of the dog. Circ Res 1974; 35:33.
96. Bishop YMM, Fienberg SE, Holland PW. Discrete Multivariate Analysis. 6th ed. Cambridge: MIT Press, 1980.
97. Bisognano JD, Weiberger HD, Bohmeyer TS et al. Myocardial-direct overexpression of the human b1-adrenergic receptor in transgenic mice. J Mol Cell Cardiol 2000; 32:817.
98. Black A, Black MM, Gensini G et al. Exertion and acute coronary artery injury. Circulation 1965; 32(Suppl 2):3.
99. Blake HA, Manion WC, Mattingly TW et al. Coronary artery anomalies. Circulation 1964; 30:927.
100. Blautass AH. Production of ventricular hypertrophy simulating idiopathic hypertrophic subaortic stenosis (IHSS) by subhypertensive infusion of norepinephrine (NE) in teh conscious dog (abstract). Clin Res 1975; 23:77.
101. Bloom S, Cancilla DA. Myocytolysis and mitochondrial calcification in rat myocardium after low doses of isoproterenol. Am J Pathol 1969; 54:373.
102. Blumgart HL, Schlesinger MJ, Davis D. Studies on the relation of the clinical manifestations of angina pectoris, coronary thrombosis and myocardial infarction to the pathologic findings with particular reference to the significance of the collateral circulation. Am Heart J 1940; 19:1.

103. Bodenheimer MM, Banka VS, Hermann GA et al. The effect of severity of coronary artery obstructive disease and the coronary collateral circulation on local histopathologic and electrographic observations in man. Am J Med 1977; 63:193.
104. Bodenheimer MM, Banka VS, Hermann GA et al.Reversible asynergy: histopathologic and electrographic correlations in patients with coronary artery disease. Circulation 1976; 53:792.
105. Bolli R, Zhu W, Thornby JI et al. Time course and determinants of recovery of function after reversible ischemia in conscious dogs. Am J Physiol 1988; 254:H102.
106. Bonow RO, Quyyumi AA, Panza JA et al. Do repeated episodes of reversible myocardial ischemia lead to progressive left ventricular dysfunction in mildly symptomatic patients with coronary artery disease? J Am Coll Cardiol 1990; 15:203A.
107. Borgers M, Thoné F, Wouters L et al. Structural correlates of regional myocardial dysfunction in patients with critical coronary stenosis: chronic hibernation? Cardiovasc Pathol 1993; 2:237.
108. Bork K. Ueber die Kranzadersklerose. Virchow Arch Path Anat 1926; 262:646.
109. Born GUR, Honour AJ, Mitchell JRA. Inhibition by adenosine and by 2-chloroadenosine of the formation and embolization of platelet thrombi. Nature 1964; 4934:761.
110. Bouchardy B, Majno G. A new approach to the histologic diagnosis of early myocardial infarcts. Cardiology 1971-72; 56:327.
111. Boyd W. In: A Textbook of Pathology: Structure and Function In Disease. 7th ed. Philadelphia: Lea & Febiger 1961; 196.
112. Bradbury S. Thirty years after ligation of the anterior descending branch of the left coronary artery. Am Heart J 1942; 24:562.
113. Brain MC, Dacie JV, Hourihane DOB. Microangiopathic haemolytic anaemia: the possible role of vascular lesions in pathogenesis. Brit J Haemat 1962; 8:358.
114. Brandenburg RO. Cardiomyopathies and their role in sudden death. J Am Coll Cardiol 1985; 5:185B.
115. Branwood AW, Montgomery GL. Observation son the morbid anatomy of coronary artery disease. Scottish M.J 1956; 1:376.
116. Braunwald E, Kloner RA. The stunned myocardium: prolonged, postischemic ventricular dysfunction. Circulation 1982; 66:1146.
117. Braunwald E. Shattuck lecture. Cardiovascular medicine at the turn of the millenium: triumphs concerns and opportunities. N Engl J Med 1997; 337:1360.
118. Braunwald E. Textbook of Cardiology. 3rd ed. Philadelphia: Saunders, 1998:1225.
119. Brechenmacher C, Coumel P, Fauchier JP et al. De Subitaneis Mortibus. XXII. Intractable paroxysmal tachycardias which proved fatal in type A Wolff-Parkinson-White Syndrome. Circulation 1977; 55:408.
120. Brechenmacher C, Coumel P, James TN. De Subitaneis Mortibus. XVI. Intractable tachycardia in infancy. Circulation 1976; 53:377.
121. Bridges JM, Weaver JA. Platelet stickness. Am Heart J 1966; 71:291.
122. Brody WR, Angeli WW, Kosek JC. Histologic fate of the venous coronary artery bypass in dog. Am J Path 1972; 66:111.
123. Brouardel F, Benham FL. Death and Sudden Death. New York: William Wood Co. 1902.
124. Brown G, Alberts JJ, Fisher MM et al. Regression of coronary artery disease as a result of intensive lipid lowering therapy in men with high level of apoproteins B. N Engl J Med 1990; 323:1289.
125. Brown GB, Gallery CA, Badger RS et al. Incomplete lysis of thrombus in the moderate underlying atherosclerotic lesion during intracoronary infusion of streptokinase for acute myocardial infarction: quantitative angiographic observations. Circulation 1986; 73:653.
126. Buchner F. Patologische Anatomie der Herzinsuffizienz Verh Deutsch Ges Kreislaufforsch 1950; 16:26.
127. Buffon A, Santini SA, Ramazzotti V et al. Large sustained lipid peroxidation and reduced antioxidant capacity in the coronary circulation after brief episodes of myocardial ischemia. J Am Coll Cardiol 2000; 35:633.
128. Buja ML, Poliner LR, Parkey RW et al. Clinicopathologic study of persistently positive technetium-99 stannous pyrophosphate myocardial scintigrams and myocytolytic degeneration after myocardial infarction. Circulation 1977; 56:1016.

129. Buja ML, Willerson JT. Clinicopathologic correlates of acute ischemic heart disease syndromes. Am J Cardiol 1981; 47:343.
130. Buja ML, Willerson JT. Role of inflammation in coronary plaque disruption. Circulation 1994; 89:503.
131. Bulkley BH, Hutchins GM. Accelerated "atherosclerosis". A morphologic study of 97 saphenous vein coronary artery bypass grafts. Circulation 1977; 55:163.
132. Bulkley BH, Klacsmann PG, Hutchins GM. Angina pectoris, myocardial infarction and sudden cardiac death with normal coronary arteries: a clinicopathologic study of 9 patients with progressive systemic sclerosis. Am Heart J 1978; 95:563.
133. Burch GE, De Pasquale NP. Arteriosclerosis in high pressure and low pressure coronary arteries. Am Heart J 1962; 63:720.
134. Burke AP, Farb A, Malcolm GT et al. Coronary risk factors and plaque morphology in men with coronary disease who died suddenly. N Engl J Med 1997; 336:1276.
135. Burke AP, Farb A, Malcolm GT et al. Plaque rupture and sudden death related to exertion in men with coronary artery disease. JAMA 1999; 281:921.
136. Burke AP, Subramanian R, Smialek J, Virmani R. Nonatherosclerotic narrowing of the atrioventricular node artery and sudden death. J Am Coll Cardiol 1993; 21:117.
137. Burton AC. Relation of structure to function of the tissues of the wall of blood vessels. Physiol Rev 1954; 34:619.
138. Cabin HS, Clubb KS, Vita N et al. Regional dysfunction by equilibrium radionuclide angiocardiography: a clinicopathologic study evaluating the relation of degree of dysfunction to the presence and extent of myocardial infarction. J Am Coll Cardiol 1987; 4:743.
139. Calif RM, Abdelmeguid AE, Kuntz RE et al. Myonecrosis after revascularization procedures. J Am Coll Cardiol 1998; 31:241.
140. Califf RM, McKinnis RA, Burks J et al. Prognostic implications of ventricular arrhythmias during 24 hour ambulatory monitoring in patients undergoing cardiac catheterization for coronary artery disease. Am J Cardiol 1982; 50:23.
141. Calkins H, Allman K, Bolling S et al. Correlation between scintigraphic evidence of regional sympathetic neuronal dysfunction and ventricular refractoriness in the human heart. Circulation 1993; 89:172.
142. Cannon WB. "Woodoo" death. Am Anthropol 1942; 44:169.
143. Cao JM, Fishbein MC, Han JB et al. Relationship between regional cardiac hyperinnervation and ventricular arrhythmia. Circulation 2000; 101:1960.
144. Carleton RA, Boyd T. Traumatic laceration of the anterior coronary artery treated by ligation without myocardial infarction: report of a case with a review of the literature. Am Heart J 1955; 56:136.
145. Carlson RE, Schott RJ, Buda AJ. Neutrophil depletion fails to modify myocardial no reflow and functional recovery after coronary reperfusion. J Am Coll Cardiol 1989; 7:1803.
146. Carroll SM, White FC, Roth DM et al. Heparin accelerates coronary collateral development in a porcine model of coronary artery occlusion. Circulation 1993; 88:198.
147. Cebelin MS, Hirsch CS. Human stress cardiomyopathy. Myocardial lesions in victims of homicidal assaults without internal injuries. Hum Path 1980; 11:123.
148. Chandler AB, Chapman I, Erhardt LR et al.Coronary thrombosis in myocardial infarction. Report of a workshop on the role of coronary thrombosis in the pathogenesis of acute myocardial infarction. Am J Cardiol 1974; 34:823.
149. Chapman I. The cause effect relationship between recent coronary artery occlusion and acute myocardial infarction. Am Heart J 1974; 87:267.
150. Cheitlin MD, De Castro CM, McAllister HA. Sudden death as a complication of anomalous left coronary origin from the anterior sinus of Valsalva. A not-so-minor congenital anomaly. Circulation 1974; 50:780.
151. Chen C, Ma L, Linfert DR et al. Myocardial cell death and apoptosis in hibernating myocardium. J Am Coll Cardiol 1997; 30:1407.
152. Cheng TO, Bashour T, Shing BK et al. Myocardial infarction in the absence of coronary arteriosclerosis. Am J Cardiol 1972; 30:680.

153. Cheng W, Li B, Kajstura L et al. Stretch-induced programmed myocyte cell death. J Clin Invest 1995; 96:2247.
154. Chesler E, King RA, Edwards JE. The myxomatous mitral valve and sudden death. Circulation 1983; 67:632.
155. Chilian WM, Mass HJ, Williams SM et al. Microvascular occlusions promote coronary collateral growth. Am J Physiol 1990; 258 (Heart Circ Physiol 27) H1103.
156. Ciuffo AA, Ouyang P, Becker LC et al. Reduction of sympathetic inotropic response after ischemia in dogs. Contributor to stunned myocardium. J Clin Invest 1985; 75:1504.
157. Claudon DG, Claudon DB, Edwards JE. Primary dissecting aneurysm of coronary artery. A cause of acute myocardial ischemia. Circulation 1972; 45:259.
158. Cliff WJ, Heathcote CR, Moss NS et al. The coronary arteries in cases of cardiac and non cardiac sudden death. Am J Pathol 1988; 192:319.
159. Clusin WT, Bristow MR, Karagueuzian HS et al. Do calcium-dependent ionic currents mediate ischemic ventricular fibrillation? Am J Cardiol 1982; 49:606.
160. Cobb LA, Baum RS, Schaffer WA. Resuscitation from out-of-hospital ventricular fibrillation: 4 years follow-up. Circulation 1975; 51-52(Suppl 3):223.
161. Cobb LA, Werner JA, Trobaugh GB. Sudden cardiac death. I A decade's experience with out-of-hospital resuscitation. Modern Concepts Cardiov Dis 1980; 6:31.
162. Cobb LA, Werner JA, Trobaugh GB. Sudden cardiac death. II Outcome of resuscitation, management and future directions. Modern Concepts Cardiov Dis 1980; 7:37.
163. Cobb LA, Werner JA. Predictors and prevention of sudden cardiac death. In: Hurst JW ed. The Heart. 6th ed. New York: Mc Graw Hill Book Co., 1986:538.
164. Cohen MV. The functional value of coronary collaterals in myocardial ischemia and therapeutic approach to enhance collateral flow. Am Heart J 1978; 95:396.
165. Cohn JN, Ferrari R, Sharpe N et al. Cardiac remodeling. Concepts and clinical implications: a consensus paper from an international forum on cardiac remodeling. J Am Coll Cardiol 2000; 35:569.
166. Cohn PF, Maddox DE, Holman BL et al. Effect of coronary collateral vessels on regional myocardial blood flow in patients with coronary artery disease. Relation of collateral circulation to vasodilator reserve and left ventricular function. Am J Cardiol 1980; 46:359.
167. Cohn PF. Silent myocardial ischemia and infarction. 2nd ed. New York: Marcel Dekker Inc., 1989.
168. Cohnheim J, Schulthess-Rechberg von A. Ueber die Folgen der Kranzarterienverschliessung fur das Herz. Virchow Arch Path Anat 1881; 85:503.
169. Colucci WS. Apoptosis in the heart. N Engl J Med 1996; 335:1224.
170. Communal C, Singh K, Pimentel DR et al. Norepinephrine stimulates apoptosis in adult rat ventricular myocytes by activation of the b-adrenergic pathway. Circulation 1998; 98:1329.
171. Connor RCR. Focal myocytolysis and fuchsinophilic degeneration of the myocardium of patients dying with various brain lesions. Ann NY Acad Sci 1969; 156:261.
172. Constantinides P. The role of endothelial injury in arterial thrombosis and atherogenesis. In: Halonen PI, Louhita A, eds. Thrombosis and Coronary Heart Disease. Advances in Cardiology. Basel: Karger. 1970:4:67.
173. Cooley DA, Bloodwell RD, Hallman GL et al. Human cardiac transplantation. Circulation 1969; 39 (Suppl I):1.
174. Corbalan R, Verrier R, Lown B. Psychological stress and ventricular arrhythmias during myocardial infarction in the conscious dog. Am J Cardiol 1974; 34:692.
175. Corday E, Heng MK, Meerbaum S et al. Derangements of myocardial metabolism preceding onset of ventricular fibrillation after coronary occlusion. Am J Cardiol 1977; 39:880.
176. Cowley MJ, Disciascio G, Rehr RB et al. Angiographic observations and clinical relevance of coronary thrombus in unstable angina pectoris. Am J Cardiol 1989; 63:108E.
177. Cowley MJ, Hastillo A, Vetrovec GW et al. Effects of intracoronary streptokinase in acute myocardial infarction. Am Heart J 1981; 102:1149.
178. Crawford T, Dexter D, Teare RD. Coronary-artery pathology in sudden death from myocardial ischemia. Lancet: 1961; Jan 28,181.

179. Cribier A, Korsatz L, Koning R et al. Improved myocardial ischemic response and enhanced collateral circulation with long repetitive coronary occlusion during angioplasty: a prospective study. J Am Coll Cardiol 1992; 20:578.
180. Cruickshank JM, Pennert K, Sornan AE et al. Low mortality from all causes including myocardial infarction, in well-controlled hypertensives treated with a beta-blocker plus other antihypertensives. J Hypertension 1987; 5:489.
181. D'Amore PA, Thompson RW. Mechanisms of angiogenesis. Ann Rev Physiol 1987; 49:453.
182. Dae MW, Herre JM, O'Connel WJ et al. Scintigraphic assessment of sympathetic innervation after transmural versus nontransmural myocardial infarction. J Am Coll Cardiol 1991; 17:1416.
183. Dae MW, Lee RJ, Ursell PC et al. Heterogeneous sympathetic innervation in Germans shepherd dogs with inherited ventricular arrhythmia and sudden cardiac death. Circulation 1997; 63:1337.
184. Dauchin N, Vaur L, Genes N et al. Treatment of acute myocardial infarction by primary coronary angioplasty or intravenous thrombolysis in the "real world". One-year results from a nationwide French survey. Circulation 1999; 99:2639.
185. Davies MJ, Thomas A. Thrombosis and acute coronary-artery lesions in sudden cardiac ischemic death. N Engl J Med 1984; 310:1137.
186. Davies MJ, Thomas AC, Knapman PA et al. Intramyocardial platelet aggregation in patients with unstable angina suffering sudden ischemic cardiac death. Circulation 1986; 73:418.
187. Davies MJ, Woolf N, Robertson WB. Pathology of acute myocardial infarction with particular reference to occlusive coronary thrombi. Br Heart J 1976; 38:659.
188. De Marco T, Dae M, Yuen-Green MSF et al. Iodine-123 metaiodobenzylguanidine scintigraphic assessment of the transplanted human heart: evidence for late innervation. J Am Coll Cardiol 1995; 25:927.
189. De Maria R, Parodi O, Baroldi G et al. Morphological basis for thallium-201 uptake in cardiac imaging and correlates with myocardial blood flow. Eur Heart J 1996; 17:951.
190. Detrano RC, Wong ND, Doherty JM et al. Coronary calcium does not accurately predict near-term future coronary events in high-risk adults. Circulation 1999; 99:2633.
191. Deutsch E, Berger M, Kussmaul WG et al. Adaptation to ischemia during percutaneous transluminal coronary angioplasty. Clinical, hemodynamic and metabolic features. Circulation 1990; 82:2044.
192. DeWood MA, Spores J, Notske R et al. Prevalence of total coronary occlusion during the early hours of transmural myocardial infarction'. N Engl J Med 1980; 303:897.
193. DeWood MA, Stifter WF, Simpson CS et al. Coronary arteriographic findings soon after non-Q-wave myocardial infarction. N Engl J Med 1986; 315:417.
194. Dhurandhar RW, Watt DL, Silver MD et al. Printzmetal's variant form of angina with arteriographic evidence of coronary arterial spasm. Am J Cardiol 1972; 30:902.
195. Di Carli MF, Bianco-Batles M, Lauda ME et al. Effects of neuropathy on coronary blood flow in patients with diabetes mellitus. Circulation 1999; 100:813.
196. Dietz WA, Tobis JM, Isner JM. Failure of angiography to accurately depict the extent of coronary artery narrowing in three fatal cases of percutaneous transluminal coronary angioplasty. J Am Coll Cardiol 1992; 19:1261.
197. Dispersyn GD, Ausma J, Thone F et al. Cardiomyocyte remodeling during myocardial hibernation and atrial fibrillation: prelude to apoptosis. Cardiov Res 1999; 43:947.
198. Dock W. The predilection of atherosclerosis for the coronary arteries. JAMA 1946; 131:875.
199. Domburg RT, Meeterk, Berkel DFM et al. Smoking cessation reduces mortality after coronary artery bypass surgery. A 20-year follow-up study. J Am Coll Cardiol 2000; 36:878.
200. Donald DE. Myocardial performance after excision of the extrinsic nerves in the dog. Circ Res 1974; 34:417.
201. Dreyer WJ, Michael LH, West MS et al. Neutrophil accumulation in ischemic canine myocardium: insights into the time course, distribution and mechanism of localization during early reperfusion. Circulation 1991; 84:400.
202. Du X, Cox HS, Dart AM, Esler MD. Sympathetic activation triggers ventricular arrhythmias in rat heart with chronic infarction and failure. Cardiov Res 1999; 43:919.
203. Duren DR, Becker AE. Focal myocytolysis mimicking the electrocardiographic pattern of transmural anteroseptal myocardial infarction. Chest 1976; 4:506.

204. Ehrich W, De la Chapelle C, Cohn AE. Anatomical ontogeny (B)Man. A study of the coronary arteries. Am J Anat 1931; 49:241.
205. Eisenberg S. Blood viscosity and fibrinogen concentration following cerebral infarction. Circulation 1966; 33-34 (Suppl 2):10.
206. El-Maraghi N, Genton E. The relevance of platelet and fibrin thromboembolism of the coronary microcirculation, with special reference to sudden cardiac death. Circulation 1980; 62:936.
207. Eliot RS, Baroldi G, Leone A. Necroscopy studies in myocardial infarction with minimal or no coronary luminal reduction due to atherosclerosis. Circulation 1974; 49:1127.
208. Eliot RS, Buell JC. Role of emotions and stress in the genesis of sudden death. J Am Coll Cardiol 1985; 5:95B.
209. Eliot RS. From stress to strength. New York: Bantam Books, 1994.
210. Ellis SG, Henschke CI, Sandor T et al. Time course of functional and biochemical recovery of myocardium salvaged by reperfusion. J Am Coll Cardiol 1983; 1:1047.
211. Elsasser A, Schlepper M, Klovekorn WP et al. Hibernating myocardium. An incomplete adaptation to ischemia. Circulation 1997; 96:2920.
212. Emberson JW, Muir AR. Changes in ultrastructure of rat myocardium induced by hypokalaemia. J Exp Physiol 1969; 54:36.
213. Emdin M, Marin Neto A, Carpeggiani C et al. Heart rate variability and cardiac denervation in Chagas' disease. J Ambul Monit 1992; 5:251.
214. Engel G. Sudden and rapid death during psychological stress. Folklore or folk wisdom. Ann Int Med 1971; 74:771.
215. Engler RL, Dahljren MD, Morris DD et al. Role of leukocytes in response to acute myocardial ischemia and reflow in dogs. Am J Physiol 1986; 251:H314.
216. Entman ML, Ballantyne CM. Inflammation in acute coronary syndromes. Circulation 1993; 88:800.
217. Entman ML, Hackel DB, Martin AM et al. Prevention of myocardial lesions during hemorrhagic shock in dogs by pronethalol. Arch Pathol 1965; 83:392.
218. Entman ML, Michael LH, Rossen RD et al. Inflammation in the course of early myocardial ischemia. Faseb J 1991; 5:2529.
219. Ericson K, Saldeen TGP, Lindquist O et al. Relationship of Chlamydia pneumoniae infection to severity of human coronary atherosclerosis. Circulation 2000; 101:2568.
220. Escaned J, Suylen van RJ, MacLeod DC et al. Histologic characteristic of tissue excised during directional coronary atherectomy in stable and unstable angina pectoris. Am J Cardiol 1993; 71:1442.
221. Estanol BV, Loyo MV, Mateos HJ et al. Cardiac arrhythmias in experimental subarachnoid hemorrhage. Stroke 1977; 8:440.
222. Esterly JA, Glacow S, Ferguson DJ. Morphogenesis of intimal obliterative hyperplasia of small arteries in experimental pulmonary hypertension. An ultrastructural study of the role of smooth-muscle cells. Am J Path 1968; 52:325.
223. Factor SM, Minase T, Cho S et al. Microvascular spasm in the cardiomyopathic syrian hamster: a preventable cause of focal myocardial necrosis. Circulation 1982; 66:342.
224. Factor SM. Smooth muscle contraction bands in the media of coronary arteries: a postmortem marker of antemortem coronary spasm. J Am Coll Cardiol 1985; 6:1326.
225. Falk E. Unstable angina with fatal outcome: dynamic coronary thrombosis leading to infarction and/or sudden death. Autopsy evidence of recurrent mural thrombosis with peripheral embolization culminating in total vascular occlusion. Circulation 1985; 71:699.
226. Falk E. Why do plaques rupture? Circulation 1992; 86 (Suppl 3):3.
227. Falk RH, Rubinow A, Cohen AS. Cardiac arrhythmias in systemic amyloidosis: correlation with echocardiographic abnormalities. J Am Coll Cardiol 1984; 3:107.
228. Fangman RJ, Hellwig CA. Histology of coronary arteries in newborn infants. Am J Path 1947; 23:901.
229. Farb A, Kolodgie FD, Jenkins M, Virmani R. Myocardial infarct extension during reperfusion after coronary artery occlusion: pathologic evidence. J Am Coll Cardiol 1993; 21:1245.
230. Farb A, Tang AL, Burke Al et al. Sudden coronary death. Frequency of active coronary lesions, inactive coronary lesions and myocardial infarction. Circulation 1995; 92:1701.
231. Faruqui AMA, Maloy WC, Felner JM et al. Symptomatic, myocardial bridging of coronary artery. Am J Cardiol 1978; 41:1305.

232. Fedele FA, Gewirtz H, Capone RJ et al. Metabolic response to prolonged reduction of myocardial blood flow distal to a severe coronary artery stenosis. Circulation 1988; 78:729.
233. Feldman AM, Combes A, Wagner D et al. The role of tumor necrosis factor in the pathophysiology of heart failure. J Am Coll Cardiol 2000; 35:537.
234. Feldman LJ, Himbert D, Juliard JM et al. Reperfusion syndrome: relationship of coronary blood flow reserve to left ventricular function and infarct size. J Am Coll Cardiol 2000; 35:1162.
235. Ferrans VJ, Buja M, Maron BJ. Myofibrillar abnormalities following cardiac muscle cell injury. In: Fleckstein A, Rona G, eds. Recent Advances in Studies on Cardiac Structure and Metabolism. Vol 6. Baltimore: University Park Press, 1975:367.
236. Ferrari R, Alfieri O, Curello S et al. Occurrence of oxidative stress during reperfusion of the human heart. Circulation 1990; 8:201.
237. Feyter de PJ, Ozaki Y, Baptista J et al. Ischemia-related lesion characteristics in patients with stable or unstable angina. A study with intracoronary angioscopy and ultrasound. Circulation 1995; 92:1408.
238. Fineschi V, Agricola E, Baroldi G et al. Myocardial morphology of acute carbon monoxide toxicity: a human and experimental morphometric study. Int J Legal Med 2000; 113:262.
239. Fineschi V, Baroldi G, Centini F et al. Cardiac oxidative stress to intraperitoneal cocaine exposure and morphologic markers indicative of myocardial injury: an experimental study in rats. Int J Legal Med in press.
240. Fineschi V, Baroldi G, Manciotti F. Anabolic steroid abuse and cardiac sudden death: a pathologic report of two cases. Arch Path Lab Med 2001; 2:253.
241. Fineschi V, Paglicci Reatelli L, Baroldi G. Coronary artery aneursysm in a yound adult: a case of sudden death. A late sequelae of Kawasaki disease. Int J Legal Med 1999; 112:120.
242. Fineschi V, Welti EV, Di Paolo M et al. Myocardial necrosis and cocaine. A quantitative morphologic study in 26 cocaine-associated deaths. Int J Leg Med 1997; 3:164.
243. Fisch C, Armstrong WP, Bigger TJ et al. Sudden Cardiac Death. Clinical electrophysiology and electrocardiography: Summary. J Am Coll Cardiol 1985; 5 (Suppl 6):27B.
244. Fishbein MC, Siegel RJ. How big are coronary atherosclerotic plaques that rupture? Circulation 1996; 94:2662.
245. Flaherty JT, Pierce JE, Ferrans VJ et al. Endothelial nuclear pattern in the canine arterial tree with particular reference to hemodynamic events. Circul Res 1972; 30:23.
246. Flameng W, Schwarz F, Hehrlein FW. Intraoperative evaluation of the functional significance of coronary collateral vessels in patients with coronary artery disease. Am J Cardiol 1978; 42:187.
247. Flameng W, Van Belle H, Vanhaecke J et al. Relation between coronary artery stenosis and myocardial purine metabolism, histology and regional function in humans. J Am Coll Cardiol 1987; 9:1235.
248. Fleckenstein A, Janke J, Doring HJ et al. Myocardial fiber necrosis due to intracellular Ca overload. A new principle in cardiac pathophysiology. In Recent Advances in Studies on Cardiac Structure and Metabolism. Dhallas NS ed, Vol 4, University Park Press, Baltimore,1975, p 563.
249. Fleisch M, Billinger M, Eberly FR et al. Physiologically assessed coronary collateral flow and intracoronary growth factor concentration in patients with 1 to 3 vessel coronary disease. Circulation 1999; 100:943.
250. Flesh M, Maack C, Cremers B et al. Effect of b-blockers on free radical-induced cardiac contractile dysfunction. Circulation 1999; 100:346.
251. Flynn MS, Kern MJ, Donohue TJ et al. Alterations of coronary collateral blood flow velocity during intraaortic balloon pumping. Am J Cardiol 1993; 71:1451.
252. Folts JD, Gallagher K, Rowe GG. Blood flow reduction in stenosed canine coronary arteries: vasospasm or platelet aggregation? Circulation 1982; 65:248.
253. Forman MB, Collins W, Kopelman HA et al. Determinants of left ventricular aneurysm formation after anterior myocardial infarction: a clinical and angiographic study. J Am Coll Cardiol 1986; 8:1256.
254. Forman MD, Oates JA, Robertson D et al. Increased adventitial mast cells in a patient with coronary spasm. N Engl J Med 1985; 313(18):1138.
255. Forrester JS, Litvack F, Grundfest W et al. A perspective of coronary disease seen through the arteries of living man. Circulation 1987; 75:505.

256. Fozzard HA. Electromechanical dissociation and its possible role in sudden cardiac death. J Am Coll Cardiol 1985; 5(Suppl 6):31B.
257. Freasure-Smith N, Lesperance F, Gravel G et al. Social support, depression and mortality during the first year after myocardial infarction. Circulation 2000; 101:1919.
258. Freifeld AG, Schuster EH, Bulkley BH. Non transmural versus transmural myocardial infarction. A morphological study. Am J Med 1983; 75:423.
259. French JE, Jennings MA, Poole JCF et al. Intimal changes in the arteries of aging swine. Proc Roy Soc 1962; 158:24.
260. Friedman M, Bovenkamp GJ. The pathogenesis of coronary thrombus. Am J Path 1966; 48:19.
261. Friedman M, Manwaring JH, Rosenman RH et al. Instantaneous and sudden deaths. Clinical and pathological differentation in coronary artery disease. JAMA 1973; 225:1319.
262. Friedman M. The coronary canalized thrombus. Provenance, structure function and relationship to death due to coronary artery disease. Brit J Exp Path 1967; 48:556.
263. Friedman M. The pathogenesis of coronary plaques, thromboses and hemorrhages: an evaluative review. Circulation 1975; 51-52(Suppl 3):34.
264. Frink RJ, Trowbridge JO, Roney PA Jr. Nonobstructive coronary thrombosis in sudden cardiac death. Am J Cardiol 1978; 42:48.
265. Fry DL. Certain chemorheologic considerations regarding the blood vascular interface with particular reference to coronary artery disease. Circulation 1969; 40 (Suppl 4):38.
266. Fujita M, Sasayama S, Asanoi H et al. Improvement of treadmil capacity and collateral circulation as a result of exercise with heparin pretreatment in patients with effort angina. Circulation 1988; 5:1022.
267. Fujiwara H, Onodera T, Tanaka M et al. A clinicopathologic study of patients with hemorrhagic myocardial infarction treated with selective thrombolysis with urokinase. Circulation 1986; 73:749.
268. Fulton M, Julian DG, Oliver MF. Sudden death and myocardial infarction. Circulation 1969; 40 (Suppl. 4):182.
269. Fulton WFW. The Coronary Arteries, Microanatomy and Pathogenesis of Obliterative Coronary Artery Disease. Springfield: Charles C. Thomas Publ.,1965
270. Fung AY, Rabkin SW. Beneficial effects of streptokinase on left ventricular function after myocardial reoxygenation and reperfusion following global ischemia in the isolated rabbit heart. J Cardiovssc Pharmacol 1984; 6:429.
271. Fuster V, Badimon JJ, Badimon L. Clinical-pathological correlations of coronary disease progression and regression. Circulation 1992; 86 (Suppl 3):31.
272. Galassi AR, Crea F, Aranjo LI et al. Comparison of regional myocardial blood flow in syndrome X and one-vessel coronary artery disease. Am J Cardiol 1993; 72:134.
273. Ganz W, Buchbinder N, Marcus H et al. Intracoronary thrombolysis in acute myocardial infarction: experimental background and clinical experience. Am Heart J 1981; 102:1145.
274. Gavin JB. The no-reflow phenomenon in ischemic myocardium. Int Rev Exp Path 1983; 25:361.
275. Geer JC, McGill HC, Robertson WB et al. Histologic characteristics of coronary artery fatty streaks. Lab Invest 1968; 18:565.
276. Geft IL, Fishbein MC, Ninomiya K et al. Intermittent brief periods of ischemia have a cumulative effect and may cause myocardial necrosis. Circulation 1982; 66:1150.
277. Geiringer E. Intimal vascularization and atherosclerosis. J Path Bact 1951; 63:201.
278. Geiringer E. The mural coronary. Am Heart J 1951a; 41:359.
279. Geng YJ, Libby P. Evidence for apoptosis in advanced human atheroma: colocalization with interleukin-1b-converting enzyme. Am J Pathol 1995; 147:251.
280. Gensini GG, Costa BCB. The coronary collateral circulation in living man. Am J Cardiol 1969; 24:393.
281. Gensini GG, Kelly AE. Incidence and progression of coronary artery disease. An angiographic correlation in 1263 patients. Arch Intern Med 1972; 129:814.
282. Gerber BL, Wijns W, Vanoverschelde JJ et al. Myocardial perfusion and oxygen consumption in reperfused dysfunctional myocardium after unstable angina. J Am Coll Cardiol 1999; 34:1939.
283. Gerdes MA, Kellerman SE, Moore AJ et al. Structural remodeling of cardiac myocytes in patients with ischemic cardiomyopathy. Circulation 1992; 86:426.

284. Ghidoni JJ, Liotta D, Thomas H. Massive subendocardial damage accompanying prolonged ventricular fibrillation. Am J Path 1969; 56:15.
285. Gibbons LW, Cooper KH, Meyer BM et al. The acute cardiac risk of strenuous exercise. JAMA 1980; 244:1799.
286. Gibson MC, Ryan KA, Murphy SA et al. Impaired coronary blood flow in nonculprit arteries in the setting of acute myocardial infarction. J Am Coll Cardiol 1999,34:974.
287. Glagov S, Weisemberg E, Zarnes CK et al. Compensatory enlargement of human atherosclerotic coronary arteries. N Engl J Med 1987; 316:1371.
288. Goldstein RE, Borer JS, Epstein SE. Augmentation of contractility following ischemia in the isolated supported heart. Am J Cardiol 1972; 29:265.
289. Goldstein S, Landis JR, Leighton R et al. Characteristics of the resuscitated out-of-hospital cardiac arrest victim with coronary heart disease. Circulation 1981; 64:977.
290. Goldstein S. Sudden Death and Coronary Heart Disease. New York: Futura Publ. Co. Inc., 1974.
291. Goldstein S. The necessity of a uniform definition of sudden coronary death: witnessed death within 1 hour of the onset of acute symptoms. Am Heart J 1982; 103:156.
292. Gomes AJ, Alexopoulos D, Winters SL et al. The role of silent ischemia, the arrhythmic substrate and the short-long sequence in the genesis of sudden cardiac death. J Am Coll Cardiol 1989; 14:1618.
293. Goodwin JF, Krikler DM. Arrhythmia as a cause of sudden death in hypertophic cardiomapathy. Lancet 1976 Oct 30; 2:937-40
294. Goodwin JF, Krikler DM. Sudden death in cardiomiopathy. In: Sudden Coronary Death. Manninen V, Halonen PI, eds. Karger, Basel Adv Cardiol 1978; 25:9.
295. Goodwin JF. The frontiers of cardiomyopathy. Brit Heart J 1982; 48,1.
296. Gorlin R, Fuster V, Ambrose JA. Anatomic-physiologic links between acute coronary syndromes. Circulation 1986; 74:6.
297. Gotlieb A, Masse S, Allard J et al. Concentric hemorrhagic necrosis of the myocardium. A morphological and clinical study. Hum Pathol 1977; 8:27.
298. Gotoh K, Minamino T, Katoh O et al. The role of intracoronary thrombus in unstable angina: angiographic assessment and thrombolytic therapy during ongoing anginal attacks. Circulation 1988; 77:526.
299. Gottlieb R, Burleson KO, Kloner RA et al. Reperfusion injury induces apoptosis in rabbit cardiomyocytes. J Clin Invest 1994; 94:1621.
300. Gradman AH, Bell PA, De Busk RF. Sudden death during ambulatory monitoring. Clinical and electrocardiographic correlations. Report of a case. Circulation 1977; 55:210.
301. Graham AF, Schroeder JS, Rider AK et al. Accelerated atherosclerosis in the human transplanted heart: precipitating factors and possible prevention. Circulation 1972; 46(Suppl 2):60.
302. Gravanis MB, Robinson K, Santoian EC et al. The reparative phenomena at the site of balloon angioplasty in human and experimental models. Cardiovasc Pathol 1993; 2:263.
303. Gregg DE, Patterson RE. Functional importance of the coronary collaterals 1980; 303:1404.
304. Gregg DE. Coronary Circulation in Health and Disease. Philadelphia: Lea Febiger, 1950.
305. Gregg DE. The natural history of collateral development. Circ Res 1974; 35:335.
306. Griffith LSC, Achuff SC, Conti R, at al. Changes in intrinsic coronary circulation and segmental ventricular motion after saphenous-vein coronary bypass graft surgery. New Engl J Med 1973; 288:590.
307. Gropel RJ, Siegel BA, Perez JE. Recovery of contractile function in viable but dysfunctional myocardium is dependent upon maintenance of oxidative metabolism. J Am Coll Cardiol 1990, 15:203A.
308. Guerra S, Leri A, Wang X et al. Myocyte death in the failing human heart is gender dependent. Circ Res 1999; 85:856.
309. Guthrie RB, Vlodaver Z, Nicoloff DM, Edwards JE. Pathology of stable and unstable angina pectoris. Circulation 1975; 51:1059.
310. Habib GB, Heibig J, Forman SA et al. Influence of coronary collateral vessels on myocardial infarct size in humans. Results of phase I thrombolysis in myocardial infarction (TIMI) trial. Circulation 1991; 83:739.
311. Hackel DB, Reimer KA. Sudden death cardiac and other causes. Durham: Carolina Academic Press, 1993.

312. Hackett D, Verwilghen J, Davies G et al. Coronary stenoses before and after acute myocardial infarction. Am J Cardiol 1989; 63:1517.
313. Haerem JW. Myocardial lesions in sudden unexpected coronary death. Am Heart J 1975; 90:562.
314. Haft JI, Al-Zarka AM. Comparison of the natural history of irregular and smooth coronary lesions: insights into the pathogenesis, progression and prognosis of coronary atherosclerosis. Am Heart J 1993; 126:551.
315. Hagstrom RM, Billings FT, Chapnick EM et al. Sudden death in survivors of myocardial infarction. Circulation 1964; 29-30 (Suppl 3):91.
316. Halpert I, Goldberg D, Levine AB et al. Reinnervation of the transplanted human heart as evidenced from heart rate variability studies. Am J Cardiol 1996; 77:180.
317. Hamer A, Vohra J, Hunt D et al. Prediction of sudden death by electrophysiologic studies in high risk patients surviving acute myocardial infarction. Am J Cardiol 1982; 50:223.
318. Hammer A. Ein Fall von trombotischen Verschlusse einer der Kranzarterien des Herzens. Wiener Medizinische Wochenschrift 1878; 5:83.
319. Hammermeister KE, De Rouen TA, Murray JA et al. Effect of aortocoronary saphenous vein bypass grafting on death and sudden death. Comparison of nonrandomized medically and surgically treated cohorts with comparable coronary disease and left ventricular function. Am J Cardiol 1977; 39: 925.
320. Hamperl H. Zur fragmentatio miocardii. Beitr path Anat 1929; 82:597.
321. Hand DKM, Haudenshild CC, Hong MK et al. Evidence for apoptosis in human atherogenesis and in rat vascular injury model. Am J Pathol 1995; 147:267.
322. Hardaway RM, McKay DG, Hollowell OW. Vascular spasm and disseminated intravascular coagulation. Influence of the phenomena one on the other. Arch Surg 1961; 83:183.
323. Hassler O. The origin of the cell constituting arterial intima thickening. An experimental autoradiographic study with the use of $H^3$-thymidine. Lab Invest 1970; 2:286.
324. Hawkey CM. Production of cardiac muscle abnormalities in offspring of rats receiving triiodothyroacetic acid (triac) and the effect of betadrenergic blockade. Cardiov Res 1981; 15:196.
325. Hearse DJ, Humphrey SM, Boink ABTJ et al. The calcium paradox: metabolic, electrophysiological, contractail and ultrastructural characteristics in four species. Eur J Cardiol 1978; 7:241.
326. Heath D. Cardiac fibroma. Br Heart J 1969; 31:656.
327. Hecht GM, Klues HG, Roberts WC et al. Coexistence of sudden cardiac death and end-stage heart failure in familial hypertrophic cardiomyopathy. J Am Coll Cardiol 1993; 22:489.
328. Helfant RH, Kemp HG, Gorlin R. The interrelation between extent of coronary artery disease, presence of collaterals, ventriculographic abnormalities and hemodynamics. Am J Cardiol 1970; 25:102.
329. Hellerstein HK, Santiago-Stevenson D. Atrophy of the heart: a correlative study of eight-five proved cases. Circulation 1959; 1:93.
330. Hellstrom HR. The injury-spasm (ischemia-induced hemostatic vasoconstrictive) and vascular autoregolatory hypothesis of ischemic disease. Resistance vessel-spasm hypothesis of ischemic disease. Am J Cardiol 1982; 49:802.
331. Henegar JR, Schwartz DD, Janicki JS. ANG II-related myocardial damage: role of cardiac sympathetic catecholamines and beta-receptor regulation. Am J Physiol 1998; 275(2 Pt 2):H534.
332. Herdson PM, Sommers HM, Jennings RB. A comparative study of the fine structure of normal and ischemic dog myocardium with special reference to early change following temporary occlusion of a coronary artery. Am J Pathol 1975; 46:367.
333. Hermans WRM, Rensing BJ, Foley DP et al. Therapeutic dissection after succesful coronary balloon angioplasty: no influence on restenosis or on clinical outcome in 693 patients. J Am Coll Cardiol 1992; 20:767.
334. Herrick JB. Clinical features of sudden obstruction of the coronary arteries. JAMA 1912; 59:87.
335. Herrick JB. Thrombosis of the coronary arteries. JAMA 1919; 72:93.
336. Hess ML, Hastillo A, Mohanakumar DVM et al. Accelerated atherosclerosis in cardiac transplantation: role of cytotossic B-cell antibodies and hyperlipidemia. Circulation 1983; 68 (suppl 2):94.
337. Higgins CB. Coronary arteriography. Key references. Circulation 1982; 62:459.
338. Hinkle LE, Argyros DC, Hayes JC et al. Pathogenesis of an unexpected sudden death: role of early cycle ventricular premature contractions. Am J Cardiol 1977; 39:873.

339. Hirai T, Fujita M, Nakajima H et al. Importance of collateral circulation for prevention of left ventricular aneurysm formation in acute myocardial infarction. Circulation 1989; 79:791.
340. Hombach V, Hoher M, Kochs M et al. Pathophysiology of unstable angina pectoris-correlation with coronary angioscopic imaging. Eur Heart J 1988; (Suppl N):40.
341. Hood WB. Experimental myocardial infarction III Recovery of left ventricular function in the healing phase. Contribution of increased fiber shortening in non infarcted myocardium. Am Heart J 1970; 79:531.
342. Hori M, Gotoh K, Kitakaze M et al. Role of oxygen-derived free radicals in myocardial edema and ischemia in coronary microvascular embolization. Circulation 1991; 84:828.
343. Hort W. Mikroskopische Beobachtung an menschlichen Infarktherzen. Virchow Arch Path Anat 1968; 345:61.
344. Hottenrott C, Buckerg GD. Studies of the effects of ventricular fibrillation on the adequacy of regional myocardial flow. Electrical versus spontaneous fibrillation. J Thorac Cardiovasc Surg 1974; 68:615.
345. Hundey WG, Lange RA, Clark CD et al. Assessment of coronary arterial flow and flow reserve in humans with magnetic resonance imaging. Circulation 1996; 93:1502.
346. Hunt D, Gore I. Myocardial lesion following experimental intracranial hemorrhage: prevention with propranolol. Am Heart J 1972; 83:232.
347. Hurst JW. Notes on teaching how some words prevent teaching. JAMA 1967; 74:858.
348. Hutchins GM, Bulkley BH, Ridolfi RL et al. Correlation of coronary arteriograms and left ventriculograms with postmortem studies. Circulation 1977; 56:32.
349. Huttner I, More RH, Rona G. Fine structural evidence of specific mechanism for increased endothelial permeability in experimental hypertension. Am J Path 1970; 61:395.
350. Ideker RE, Behar VS, Wagner GS et al. Evaluation of asynergy as an indicator of myocardial fibrosis. Circulation 1978; 57:715.
351. Inoue H, Zipes DP. Results of sympathetic denervation in the canine heart: supersensitivity that may be arrhythmogenic. Circulation 1987; 75:877.
352. Ip JH, Fuster V, Badimon L et al. Syndromes of accelerated atherosclerosis:role of vascular injury and smooth muscle cell proliferation. J Am Coll Cardiol 1990; 15:1667.
353. Irniger W. Histologische Alterbestimmung von Thrombosen und Embolien. Virch Arch path Anat 1963; 336:220.
354. Isner JM, Kearney M, Bortman S et al. Apoptosis in human atherosclerosis and restenosis. Circulation 1995; 91:2703.
355. Jacoby RM, Nesto RW. Acute myocardial infarction in the diabetic patient: pathophysiology, clinical course and prognosis. J Am Coll Cardiol 1992; 20:736.
356. James TN, Armstrong RS, Silverman J et al. De Subitaneis Mortibus. VI. Two young soldiers. Circulation 1974; 49:1239.
357. James TN, Beeson CW II, Sherman EB et al. De Subitaneis Mortibus. XIII. Multifocal Purkinje cell tumors of the heart. Circulation 1975; 52:333.
358. James TN, Carson NAJ, Froggatt P. De Subitaneis Mortibus. IV. Coronary vessels and conduction system in homocystinuria. Circulation 1974; 49:367.
359. James TN, Derek JL, Carson DJ et al. De Subitaneis Mortibus.1. Fibroma compressing His bundle. Circulation 1973; 48:428.
360. James TN, Frame B, Coates EO. De Subitaneis Mortibus. III. Pickwickian Syndrome. Circulation 1973; 48:1311.
361. James TN, Froggatt P, Atkinsons Jr WJ et al. De Subitaneis Mortibus. XXX. Observations on the pathophysiology of the long QT syndromes with special reference to the neuropathology of the heart. Circulation 1978; 57:1221.
362. James TN, Froggatt P, Marshall TK. De Subitaneis Mortibus. II. Coronary embolism in the fetus. Circulation 1973; 48:890.
363. James TN, Galakhov I. De Subitaneis Mortibus. XXVI. Fatal electrical instability of the heart associated with benign congenital polycystic tumor of the atrioventricular node. Circulation 1977; 56:667.
364. James TN, Hackel DB, Marshall TK. De Subitaneis Mortibus. V. Occluded A-V node artery. Circulation 1974; 49:772.

365. James TN, Haubrich WS. De Subitaneis Mortibus. XIV. Bacterial arteritis in Whipple's disease. Circulation 1975; 52:722.
366. James TN, Jackson DA. De Subitaneis Mortibus. XXVII. Histological abnormalities in the sinus node, atrioventricular node and His bundle associated with coarctation of the aorta. Circulation 1977; 56:1094.
367. James TN, Marilley Jr RJ, Marriott HJ. De Subitaneis Mortibus. XI. Young girl with palpitations. Circulation 1975; 51:743.
368. James TN, Marshall ML, Craig MW. De Subitaneis Mortibus. VII. Disseminated intravascular coagulation and paroxysmal atrial tachycardia. Circulation 50:395,1974
369. James TN, Marshall TK. De Subitaneis Mortibus. XII. Asymmetrical hypertrophy of the heart. Circulation 1975; 51:1149.
370. James TN, Marshall TK. De Subitaneis Mortibus. XVII. Multifocal stenoses due to fibromuscolar dysplasia of the sinus node artery. Circulation 1976; 53:736.
371. James TN, Marshall TK. De Subitaneis Mortibus. XVIII. Persistent fetal dispersion of the atrioventricular node and His bundle within the central fibrous body. Circulation 1976; 53:1026.
372. James TN, Marshall TK, Edwards JE. De Subitaneis Mortibus. XX. Cardiac electrical instability in the presence of a left superior vena cava. Circulation 1976; 54:689.
373. James TN, McKone RC, Hudspeth AS. De Subitaneis Mortibus. X. Familial congenital heart block. Circulation 1975; 51:379.
374. James TN, Pearce WN, Givhan EG. Sudden death while driving. Role of sinus perinodal degeneration and cardiac neural degeneration and ganglionitis. Am J Cardiol 1980; 45:1095.
375. James TN, Puech P. De Subitaneis Mortibus. IX. Type A Wolff-Parkinson-White Syndrome. Circulation 1974; 50:1264.
376. James TN, Robertson BT, Waldo AL, Branch CE. De Subitaneis Mortibus. XV. Hereditary stenosis of the His bundle in pug dogs. Circulation 1975; 52:1152.
377. James TN, Schlant RC, Marshall TK. De Subitaneis Mortibus. XXIX. Randomly distributed focal myocardial lesions causing destruction in the His bundle or a narrow-origin left bundle branch. Circulation 1978; 57:816.
378. James TN, Spencer MS, Kloepfer JC. De Subitaneis Mortibus. XXI. Adult onset syncope, with comments on the nature of congenital heart block and the morphogenesis of the human atrioventricular septal junction. Circulation 1976; 54:1001.
379. James TN. De Subitaneis Mortibus. VIII. Coronary arteries and conduction system in scleroderma heart disease. Circulation 1974; 50:844.
380. James TN. De Subitaneis Mortibus. XIX. On the cause of sudden death in pheochromocytoma, with special reference to the pulmonary arteris, the cardiac conduction system, and the aggregation of platelets. Circulation 1976; 54:348.
381. James TN. De Subitaneis Mortibus. XXIII. Rheumatoid arthritis and ankylosing spondylitis. Circulation 1977; 55:669.
382. James TN. De Subitaneis Mortibus. XXIV. Ruptured interventricular septum and heart block. Circulation 1977; 55:934.
383. James TN. De Subitaneis Mortibus. XXV. Sarcoid heart disease. Circulation 1977; 56:320.
384. James TN. De Subitaneis Mortibus. XXVIII. Apoplexy of the heart. Circulation 1978; 57:385.
385. James TN. Degenerative lesions of a coronary chemoreceptor and nearby neural elements in the heart of victims of sudden death. J Am Coll Cardiol 1986; 8:12A.
386. James TN. Morphologic substrates of sudden death: summary. J Am Coll Cardiol 1985; 5 81B.
387. James TN. Pathology of small coronary arteries. Am J Cardiol 1967; 20:679.
388. Jennings RB. Early phase of myocardial ischemic injury and infarction. Am J Cardiol 1969; 24:753.
389. Jerome KR, Vallant C, Jaggi R. The TUNEL assay in the diagnosis of graft-versus-host disease:caveats for interpretation. Pathology 2000; 32:136.
390. Johnson DW, Flemma RJ, Lepley D. Direct reconstruction of flow to small distal coronary arteries. Am J Cardiol 1970; 25:105.
391. Johson DE, Gao SZ, Schroeder JS et al. The spectrum of coronary artery pathologic findings in human cardiac allografts. J Heart Transplant 1989; 8:349.
392. Jones CE, Devous MD, Thomas JX et al. The effect of chronic cardiac denervation on infarct size following acute coronary occlusion. Am Heart J 1978; 95:738.

393. Jorgensen L, Haerem JW, Chandler BA et al. The pathology of acute coronary death. Acta Anesth Scand 1968; 29:193.
394. Jorgensen L, Hovig T, Rowsell HC et al. Adenosine-diphosphate-induced platelet aggregation and vascular injury in swine and rabbits. Am J Path 1970; 61:161.
395. Jorgensen L, Rowsell HC, Hovig T et al. Adenosine-diphosphate-induced platelet aggregation and myocardial infarction in swine. Lab Invest 1967; 17:616.
396. Jost S, Deckers JW, Nikutta P et al. Progression of coronary artery disease is dependent on anatomic location and diameter. J Am Coll Cardiol 1993; 2:1339.
397. Kahn JP, Perumal AS, Gully RJ et al. Correlation of type A behaviour with adrenergic receptor density: implications for coronary artery disease pathogenesis. Lancet 1987; 24:937.
398. Kanoh M, Takemura G, Misao J et al. Significance of myocytes with positive DNA in situ nick end-labeling (Tunel) in hearts with dilated cardiomyopathy. Not apoptosis but DNA repair. Circulation 1999; 99:2757.
399. Kaplan J. The influence of sympathetic activation and psychosocial stress on coronary artery atherosclerosis. In: Hypertension -The tip of an iceberg. Therapeutics Today Series. Manchester: Adis Press, 1988; 7:9-11.
400. Kaplan JR, Manuck SB, Adams MR, et al. The effects of betaadrenergic blocking agents on atherosclerosis and its complications. Eur Heart J 1987; 8:928.
401. Kaski JC, Crea F, Meran D et al. Local coronary supersensitivity to diverse vasoconstrictive stimuli in patients with variant angina. Circulation 1986; 74:1255.
402. Kassab GS, Imoto K, White FC et al. Coronary arterial tree remodeling in right ventricular hypertrophy. Am J Physiol 1993; 265(Heart Circ Physiol 34):H366.
403. Katz AM. The cardiomyopathy of overload: an unnatural growth response. Eur Heart J 1995; 16 (Suppl 0):110.
404. Katz AM. Cellular mechanisms in congestive heart failure. Am J Cardiol 1988; 62:3A.
405. Katz AM. Effects of ischemia on the cardiac contractile proteins. Cardiology 1971-72; 56:276.
406. Kaunitz PE. Origin of left coronary artery from pulmonary artery. Review of the literature and report of two cases. Am Heart J 1947; 33:182.
407. Kavantzas NG, Lazaris AC, Agapitos EV et al. Histologic assessment of apoptotic cell death in cardiomyopathies. Pathology 2000; 32:179.
408. Kaye DM, Lambert GW, Lefkovits J et al. Neurochemical evidence of cardiac sympathetic activation and increased central nervous system norepinephrine turnover in severe congestive heart failure. J Am Coll Cardiol 1994; 23:570-8.
409. Kelley ST, Malekan R, Gorman JH et al. Restraining infarct expansion preserves left ventricular geometry and function after acute anterolateral infarction. Circulation 1999; 99:135.
410. Kern WH, Dermer GB, Lindesmith GG. The intimal proliferation in aortic-coronary saphenous vein grafts. Am Heart J 1972; 84:771.
411. Khouri EM, Gregg DE, Lowesohn HS. Flow in the major branches of the left coronary artery during experimental coronary insufficiency in the unanesthetized dog. Circul Res 1968; 23:99.
412. Khouri EM, Gregg DE, McGranahan GM. Regression and reappearance of coronary collaterals. Am J Physiol 1971; 220:655.
413. Kini A, Marmur JD, Kini S et al. Creatine kinase-MB elevation after coronary intervention correlates with diffuse atherosclerosis and low-to-medium level elevation has a benign clinical course. J Am Coll Cardiol 1999; 34:663.
414. Kloner RA, Allen J, Zheng Y et al. Myocardial stunning following exercise treadmill testing in man. J Am Coll Cardiol 1990; 15:203A.
415. Kloner RA, Bolli R, Marban E et al. Medical and cellular implications of stunning, hibernating and preconditioning. Circulation 1998; 97:1848.
416. Kloner RA, Ganote CE, Jennings RB. The "no-reflow" phenomenon after temporary coronary occlusion in dog. J Clin Invest 1974; 54:1496.
417. Kloner RA, Przyklenk K, Patel B. Altered myocardial states. The stunned and hibernating myocardium. Am J Med 1989; 86 (Supp 1A):14.
418. Knisely MH. The settling of sludge during life. First observations, evidences and significance. A contribution to the biophysics of disease. Acta Anat 1961; 44:7.

419. Kocher AA, Schuster MD, Sgabolchs MJ et al. Neovascularization of ischemic myocardium by human bone marrow-derived angioblasts prevent cardiomyocyte apoptosis, reduces remodeling and improves cardiac function. Nature Medicine 2001; 7:430.
420. Kochi K, Takebayashi S, Hikori T et al. Significance of adventitial inflammation of the coronary artery in patients with unstable angina: Results at autopsy. Circulation 1985; 71:709.
421. Kosek JC, Chartrand C, Hurley EJ et al. Arteries in canine cardiac homografts. Ultrastructure during acute rejection. Lab Invest 1969; 21:328.
422. Kragel AH, Gertz SD, Roberts WC. Morphologic comparison of frequency and types of acute lesions in the major epicardial coronary arteries in unstable angina pectoris, sudden coronary death and acute myocardial infarction. J Am Coll Cardiol 1991; 18:801.
423. Kragel AH, Reddy SG, Wittes JT et al. Morphometric analysis of the composition of atherosclerotic plaques in the four major epicardial coronary arteries in acute myocardial infarction and sudden coronary death. Circulation 1989; 80:1747.
424. Kral BG, Becker LC, Blumenthal RS et al. Exaggerated reactivity to mental stress is associated with exercise-induced myocardial ischemia in an asymptomatic high-risk population. Circulation 1997; 96:42.
425. Kramer CM, Nicol PD, Rogers WJ et al. Reduced sympathetic innervation underlies adjacent noninfarcted region dysfunction during left ventricular remodeling. J Am Coll Cardiol 1997; 30:1079.
426. Kubler W, Strasser RH. Signal transduction in myocardial ischemia. Eur Heart J 1994; 15:437.
427. Kubler W. The sympathetic system in evolution and in ischemic heart disease: a controversy? Eur Heart J 1992; 13:1301.
428. Kuller L, Perper J, Cooper M. Demographic characteristics and trends in arteriosclerotic heart disease mortality: sudden death and myocardial infarction. Circulation 1975; 51/52 (suppl.3):1.
429. La Canna G, Alfieri O, Giubbini R et al. Echocardiography during infusion of dobutamine for identification of reversible disfunction in patients with chronic coronary artery disease. J Am Coll Cardiol 1994; 23:617.
430. Lahan RJ, Chronos NA, Pike M et al. Intracoronary basic fibroblast growth factor (FGF-2) in patients with severe ischemic heart disease: results of a phase I open-label dose escalation study. J Am Coll Cardiol 2000; 36:2132.
431. Laine P, Keartiness M, Pentilla A et al. Association between myocardial infarction and the mastcells in the adventitia of the infarct related coronary artery. Circulation 1999; 99:361.
432. Laks H, Kaiser GC, Mudd JG et al. Revascularization of the right coronary artery. Am J Cardiol 1979; 43:1109.
433. Laks MM. Myocardial hypertrophy produced by chronic infusion of sublhypertensive doses of norepinephrine in the dog. Chest 1973; 64:75.
434. Lameris TW, Zeeuw de S, Albert G et al. Time course and mechanism of myocardial catecholamine release during transient ischemia in vivo. Circulation 2000; 101:2645.
435. Lancisi GM. Opera Omnia, Tomus I De Subitaneis Mortibus, Libri duo, Tertia editione, Palladis Roma 1745.
436. Langille LB. Hemodynamic factors and vascular disease. In Silver MD ed: Cardiovascular Pathology, Churchill Livingstone, New York, 2nd Edition,1991, p.131.
437. Leachman RD, Cokkinos DV, Zamalloa O et al. Intercoronary artery steal. Cardiovasc Res Cent Bull 1972; 10:71.
438. Leary T. Coronary spasm as a possible factor in producing sudden death. Am Heart J 1935; 10:338.
439. Lee D, Eigler N, Fishbein MC et al. Identification of intracoronary thrombus and demonstration of thrombectomy by intravascular ultrasound imaging. Am J Cardiol 1994; 73:522.
440. Leenen FHH. Cardiovascular consequence of syumpathetic hyperactivity. Can J Cardiol 1999; 40(Suppl A:2A).
441. Leor J, Poole KW, Kloner RA. Sudden cardiac death triggered by an earthquake. N Engl J Med 1996; 334:413.
442. Lesch M, Kehoe RF. Predictability of sudden cardiac death: a partially fulfilled promise. N Engl J Med 1984; 310:255.
443. Levin DC, Fallon JT. Significance of the angiographic morphology of localized coronary stenoses: histopathologic correlations. Circulation 1982; 66:316.

444. Levine HD, Welles RE. Blood sludging observed at the bed side: flexibility and limitations of examination by ophthalmoscopy. Am J Cardiol 1964; 13:48.
445. Levy RL, Breunn HG. Acute coronary insufficiency. JAMA 1936; 106:1080.
446. Li ST, Tack CH, Fananapazir L et al. Myocardial perfusion and sympathetic innervation in patients with hypertrophic cardiomyopathy. J Am Coll Cardiol 2000; 35:1867.
447. Libby P. Changing concepts of atherogenesis. J Int Med 2000; 247:349.
448. Libby P. Molecular bases of the acute coronary syndromes. Circulation 1995; 91:2844.
449. Liberthson RR, Nagel EL, Hirschman JC et al. Pathophysiologic observations in prehospital ventricular fibrillation and sudden cardiac death. Circulation 1974; 49:790.
450. Lie JT, Holley KE, Kampa WR et al. New histochemical method for morphologic diagnosis of early stages of myocardial ischemia. Mayo Clin Proc 1971; 46:319.
451. Lie JT, Lawrie GM, Morris GC et al. Hemorrhagic myocardial infarction associated with aortocoronary bypass revascularization. Am Heart J 1978; 96:295.
452. Lie JT, Titus JL. Pathology of the myocardium and the conduction system in sudden coronary death. Circulation 1975; 51-52(Suppl 3):41.
453. Lie JT. Histopathology of the conduction system in sudden death from coronary heart disease. Circulation 1975; 51:446.
454. Lie JT. Sudden death from diseases of the cardiovascular system. In: Silver MD ed. Cardiovascular Pathology. 2nd ed. New York: Churchill Livingstone, 1991:811.
455. Liggett SB, Tepe NM, Lorenz JN et al. Early and delayed consequences of b-adrenergic receptor overexpression in mouse hearts. Critical role for expression level. Circulation 2000; 101:1707.
456. Likar IN, Robinson RW, Gouvelis A. Microthrombi and intimal thickening in bovine coronary arteries. Arch Path 1969; 87:146.
457. Linzbach AJ. Heart failure from the point of view of quantitative anatomy. Am J Cardiol 1960; 5:370.
458. Linzbach AJ. Mikrometrische und histologische Analyse hypertropher menschlicher Herzen. Virchows Arch Path Anat 1947; 314:534.
459. Lisa JR. Pathologic findings in the heart in sudden cardiac deaths. Ann Intern Med 1939; 12:1968.
460. Little WC, Constantinescu M, Applegate RJ et al. Can coronary angiography predict the site of a subsequent myocardial infarction in patients with mild-to-moderate coronary artery disease? Circulation 1988; 78:1157.
461. Liuzzo G, Biasucci ML, Gallimore RJ et al. The prognostic value of C-reactive protein and serum amyloid A protein in severe unstable angina. N Engl J Med 1994; 331:417.
462. Liuzzo G, Buffon A, Biasucci LM et al. Enhanced inflammatory response to coronary angioplasty in patients with severe unstable angina. Circulation 1998; 98:2370.
463. Lowe JE, Cummings RG, Adams DH et al. Evidence that ischemic cell death begins in the subendocardium independent of variations in collateral flow or wall tension. Circulation 1983; 68:190.
464. Lown B, Verrier RL, Rabinowitz SH. Neural and psychologic mechanisms and the problem of sudden cardiac death. Am J Cardiol 1977; 39:890.
465. Lown B. Sudden cardiac death: the major challenge confronting contemporary cardiology. Am J Cardiol 1979; 43:313.
466. Ludwig G. Capillary pattern of the myocardium. In: Bajusz E, Jasmin G, eds. Functional Morphology of the Heart. Meth Achievm Exp Path. Basel: Karger, 1971; 5:238.
467. Luke JL, Helpern M. Sudden unexpected death from natural causes in young adults. A review of 275 consecutive autopsied cases. Arch Path 1968; 85:10.
468. Lum M, Stevenson WG, Stevenson LW et al. Diverse mechanisms of unexpected cardiac arrest in advanced heart failure. Circulation 1989; 80:1675.
469. Luu M, Stevenson WG, Stevenson LW et al. Diverse mechanisms of unexpected cardiac arrest in advanced heart failure. Circulation 1989; 80:1675.
470. Lynch JJ, Paskewitz DA, Gimbel KS et al. Psychological aspects of cardiac arrhythmia. Am Heart J 1977; 93:645.
471. Mac Isaac AI, Thomas JD, Topol EJ. Toward the quiescent coronary plaque. J Am Coll Cardiol 1993; 22:1228.

472. Maisel AS. Beneficial effects of metaprolol treatment in congestive heart failure. Reversal of sympathetic-induced alterations of immunologic function. Circulation 1994; 90:1774.
473. Majno G, Ames A, Chaing J. No-reflow after cerebral ischemia. Lancet 1967; 2:569.
474. Majno G, Joris I. Apoptosis, oncosis and necrosis. An overview of cell death. Am J Pathol 1995; 146:3.
475. Mak T, Weiglicki WB. Protection by beta-blocking agents against free radical-mediated sarcolemmal lipid peroxidation. Circ Res 1988; 63:262.
476. Mallat Z, Tedgni A, Fontaliran F et al. Evidence of apoptosis in arrhythmogenic right ventricular dysplasia. N Eng J Med 1996; 335:1190.
477. Malliani A, Schwartz PJ, Zanchetti A. A sympathetic reflex elicited by experimental coronary occlusion. Am J Physiol 1979; 217:703.
478. Mallory GK, White PD, Salcedo-Salgar J. The speed of healing of myocardial infarction. A study of the pathologic anatomy of seventy-two cases. Am Heart J 1939; 18:647.
479. Mantini E, Tanaka S, Lillehei CW. Analysis of the earliest effects of mammary artery implantation on the ischemic ventricle. Ann Thorac Surg 1968; 5:393.
480. Manyari DE, Knudtson M, Kloiber R. Sequential Thallium-201 myocardial perfusion studies after successful percutaneous transluminal coronary artery angioplasty: delayed resolution of exercise-induced scintigraphic abnormalities. Circulation 1988; 77:86.
481. Marcus FI, Fontaine GH, Guiraudon G et al. Right ventricular dysplasia: a report of 24 adult cases. Circulation 1982; 65:384.
482. Marcus ML, Chilian WM, Kanatsuga H et al. Understanding the coronary circulation through studies at the microvascular level. Circulation 1990; 82:1.
483. Marcus ML, Kerber RE, Ehrhardt J et al. Three dimensional geometry of acutely ischemic myocardium. Circulation 1975; 52:254.
484. Margolis JR, Hirshfeld JW, McNeer FJ et al. Sudden death due to coronary artery disease. A clinical, hemodynamic and angiographic profile. Circulation 1975; 51-52(Suppl 3):180.
485. Marian AJ, Roberts R. Familial hypertrophic cardiomyopathy: a paradigm of the cardiac hypertrophic response to injury. Ann Med 1998; 30 (Suppl 1):24.
486. Marino P, Nidasio G, Golia G et al. Frequency of predischarge ventricular arrhythmias in postmyocardial infarction patients depend on residual left ventricular pump performance and is independent of the occurrence of acute reperfusion. J Am Coll Cardiol 1994; 23:290.
487. Mark DB, Naylor DC, Heatki MA et al. Use of medical resorce and quality of life after acute myocardial infarction in Canada and United States. N Engl J Med 1994; 331:1130.
488. Maron BJ, Roberts WC, Edwards JE et al. Sudden death in patients with hypertophic cardiomyopathy: characterization of 26 patients without functional limitation. Am J Cardiol 1978; 41:803.
489. Maron BJ, Roberts WC, McAllister HA et al. Sudden death in young athletes. Circulation 1980; 62:218.
490. Marron K, Wharton J, Sheppard MN et al. Distribution, morphology and neurochemistry of endocardial and epicardial nerve terminal arborizations in the human heart. Circulation 1995; 92:2343.
491. Marsch SCU, Wanigasekera VA, Ryder WA et al. Graded myocardial ischemia is associated with a decrease in diastolic distensibility of the remote nonischemic myocardium in the anesthetized dog. J Am Coll Cardiol 1993; 22:899.
492. Marti MC, Bouchardy B, Cox JN. Aorto-coronary bypass with autogenous saphenous vein grafts: histopathological aspects. Virchow Arch Path Anat 1971; 152:255.
493. Martin AM, Hackel DB. An electron microscopic study of the progression of myocardial lesions in the dog after hemorrhagic shock. Lab Invest 1966; 15:243.
494. Martin AM, Hackel DB. The myocardium of the dog in hemorrhagic shock. An istochemical study. Lab Invest 1963; 12:77.
495. Martin SJ. Apoptosis: suicide, execution or murder? Trends Cell Biol 1993; 3:141.
496. Marzilli M, Sambuceti G, Fedele S et al. Coronary microcirculatory vasoconstriction during ischemia in patients with unstable angina. J Am Coll Cardiol 2000; 35:327.
497. Maseri A, Crea F, Cianflone D. Myocardial ischemia caused by distal coronary vasoconstriction. Am J Cardiol 1992; 70:1602.

498. Maseri A, L'Abbate A, Baroldi G et al. Coronary vasospasm as a possible cause of myocardial infarction. A conclusion derived from the study of "preinfarction" angina. N Engl J Med 1978; 299:1271.
499. Mathey DG, Kuck KH, Tilsner V et al. Non surgical coronary artery recanalization in acute transmural myocardial infarction. Circulation 1981; 63:489.
500. Matsunari I, Schricke U, Bengel FM et al. Extent of cardiac sympathetic neuronal damage is determined by the area of ischemia in patients with acute coronary syndromes. Circulation 2000; 101:2579.
501. Mazzone A, De Servi S, Ricevuti G et al. Increased expression of neutrophil and monocyte adhesion molecules in unstable coronary artery disease. Circulation 1993; 88:358.
502. Mazzone A, De Servi S, Ricevuti G et al. Increased expression of neutrophil and monocyte adhesion molecules in unstable coronary artery disease. Circulation 1993; 88:358.
503. Mc Falls EO, Araujo LI, Lammertsma A et al. Vasodilator reserve in collateral-dependent myocardium as measured by positron emission tomography. Eur Heart J 1993; 14:336.
504. Mc Gill HC, Eggen DA, Guzman MA et al. The geographic pathology of atherosclerosi; Lab Invest 1968; 18:463.
505. McCance AJ, Thompson PA, Forfar JC. Increased cardiac sympathetic nervous activity in patients with unstable coronary heart disease. Eur Heart J 1993; 14:751.
506. McIntosh HD. Sudden cardiac death. Neural control of the heart: summary of discussion. J Am Coll Cardiol 1985; 5:111B.
507. McKay DG. Diseases of hypersensitivity. Disseminated intravascular coagulation. Arch Inter Med 1965; 116:83.
508. McKenna WJ, Stewart JT, Nihoyannopoulos P et al. Hypertrophic cardiomyopathy without hypertrophy: 2 families with myocardial disarray in the absence of increased myocardial mass. Br Heart J 1990; 63,287.
509. Meerson FZ, Kagan VE, Kozlov YP et al. The role of lipid peroxidation in pathogenesis of ischemic damage and the antioxidant protection of the heart. Basic Res Cardiol 1982; 77:465.
510. Meerson FZ. The myocardium in hyperfunction, hypertophy and heart failure. Circul Res 1969; 25(Suppl 2):1.
511. Mehmet CO, Roberts CS, Lemole GM. Role of lymphostasis in accelerated atherosclerosis in transplanted heart. Am J Cardiol 1987; 60:430.
512. Michele DE, Albayya FP, Metzger GM. Direct convergent hypersensitivity of calcium-activated force generation produced by hypertrophic cardiomyopathy mutant a-tropomyosine in adult cardiac monocytes. Nature Medicine 1999; 5:1413.
513. Mills JD, Fischer D, Villanueva FS. Coronary collaterals development during chronic ischemia: serial assessment using harmonic myocardial contrast echocardiography. J Am Coll Cardiol 2000; 36:618.
514. Milstein S, Buetikofer J, Lesser J et al. Cardiac asystole: a manifestation of neurally mediated hypotension-bradycardia. J Am Coll Cardiol 1989; 14:1626.
515. Minkowski WL. The coronary arteries of infants. Am J M Sc 1947; 214:623.
516. Mints GS, Painer JA, Pichard AD et al. Atherosclerosis in angiographically "normal" coronary artery reference segments: an intravascular ultrasound study with clinical correlation. J Am Coll Cardiol 1995; 25:1479.
517. Mitrani RD, Klein LS, Miles WH et al. Regional cardiac sympathetic denervation in patients with ventricular tachycardia in absence of coronary artery disease. J Am Coll Cardiol 1993; 22:1344.
518. Miyao Y, Kugiyama K, Kawano K et al. Diffuse intimal thickening of coronary arteries in patients with coronary spastic angina. J Am Coll Cardiol 2000; 36:432.
519. Mizuno K, Satomura K, Miyamoto A et al. Angioscopic evaluation of coronary artery thrombi in acute coronary syndrome. New Engl J Med 1992; 326:287.
520. Moberg A. Anastomoses between extracardiac vessels and coronary arteries. Acta Med Scand 1968; 485 (Suppl):1.
521. Moon HD. Coronary arteries in fetuses, infants and juveniles. Circulation 1957; 16:263.
522. Moore CA, Cannon J, Watson DD et al. Thallium-201 kinetics in stunned myocardium characterized by severe postischemic systolic dysfunction. Circulation 1990; 81:1622.

523. Morales AR, Romanelli R, Boucek RJ. The mural left anterior descending coronary artery, strenuous exercise and sudden death. Circulation 1980; 62:230.
524. Morales AR, Romanelli R, Tate LG et al. Intramural left anterior descending coronary artery: significance of the depth of the muscular tunnel. Hum Pathol 1993; 24:693.
525. Moreno PR, Falk E, Palacios IF et al. Macrophage infiltration in acute coronary syndromes. Implications for plaque rupture. Circulation 1994; 90:775.
526. Morgagni JB. De Sedibus et Causis Morborum per Anatomen Indagatis. Ebrodani in Elvetia,1779.
527. Morgan AD. The Pathogenesis of Coronary Occlusion. Oxford: Blackwell Scientific Publications, 1956.
528. Moritz AR, Zamcheck N. Sudden and unexpected death of young soldiers: diseases responsible for such death during world war II. Arch Path 1946; 42:459.
529. Morris JN, Gardner MJ. Epidemiology of ischemic heart disease; Am J Med 1969; 46:674.
530. Morrow DA, Rifai N, Antman EM et al. C-reactive protein is a potent predictor of mortality independently of and in combination with troponin T in acute coronary syndromes. A TIMI 11A substudy. J Am Coll Cardiol 1998; 31:1460.
531. Mortara A, La Rovere MT, Pinna GD. Nonselective beta-adrenergic blocking agent, carvedilol, improves arterial baroflex gain and heart rate variability in patients with stable chronic heart failure. J Am Coll Cardiol 2000:36:1612.
532. Moschos CB, Haider B, Khan Y et al. Relation of platelets to catecholamine induced myocardial injury. Cardiov Res 1978; 12:243.
533. Munck W. The pathological anatomy of sudden heart death. Acta Path et Microbiol Scandinav 1946; 23:107.
534. Mustard JF, Packham MA. Platelet function and myocardial infarction. Circulation 1969; 40 (Suppl 4):20.
535. Myasnikov AL, Chazov EI, Hoshevnikova TL et al. Some new data on the occurrence of coronary thrombosis in conjuction with atherosclerosis. J Atheroscl Res 1961; 1:401.
536. Nakamura M, Nishikawa H, Mukai S et al. Impact of coronary artery remodeling on clinical presentation of coronary artery disease: an intravascular ultrasound study. J Am Coll Cardiol 2001; 37:63.
537. Nakata T, Miyamoto K, Doi A et al. Cardiac death prediction and impaired cardiac sympathetic innervation assessed by MIBG in patients with failing and nonfailing hearts. J Nucl Cardiol 1998; 5:579.
538. Narkiewicz K, De Borne van PJH, Hausberg M et al. Cigarette smoking increases sympathetic outflow in humans. Circulation 1998; 98:528.
539. Narula J, Dawson MS, Singh BK et al. Noninvasive characterization of stunned, hibernating, remodeled and nonviable myocardium in ischemic cardiomyopathy. J Am Coll Cardiol 2000; 36:1913.
540. Narula J, Heider N, Virmani N et al. Apoptosis in myocytes in end-stage heart failure. N Eng J Med 1996; 335:1182.
541. Nasser FN, Walls JT, Edwards WD et al. Lidocaine-induced reduction in size of experimental myocardial infarction. Am J Cardiol 1980; 46:967.
542. Nathanson MH. Pathology and pharmacology of cardiac syncope and sudden death. Arch Intern Med 1936; 58:685.
543. Neri Serneri GG, Abbate R, Gori AM et al. Transient intermittent lymphocyte activation is responsible for the instability of angina. Circulation 1992; 86:790.
544. Neri Serneri GG, Boddi M, Arata L et al. Silent ischemia in unstable angina is related to an altered cardiac norepinephrine handling. Circulation 1993; 87:1928.
545. Nesto R, Kowalchuk G. The ischemic cascade: temporal sequence of hemodynamic, electrocardiographic and symptomatic expression of ischemia. Am J Cardiol 1987; 59:23C.
546. Newby AC, Libby P, vander Wal AC. Plaque instability: the real challange for atherosclerosis research in the next decade? Cardiov Res 1999; 41:321.
547. Newman WP, Tracy RE, Strong JP et al. Pathology of sudden coronary death. Ann NY Acad Sci 1982; 382:39.
548. Noakes TD, LH Opie, Rose AG. Autopsy-proved coronary atherosclerosis in marathon runners. N Engl J Med 1979; 301:86.

549. Noakes TD, LH Opie. Marathon running and the heart: the South African experience. Am Heart J 1979; 98:669.
550. Noble J, Bourassa MG, Peticlerc R et al. Myocardial bridging and milking effect of the left anterior descending coronary artery: normal variant or obstruction? Am J Cardiol 1997; 37:993.
551. Nohara R, Kambara H, Murakami T et al. Collateral function in early acute myocardial infarction. Am J Cardiol 1983; 52:955.
552. O'Reilly RJ, Spellberg RD. Rapid resolution of coronary arterial emboli. Myocardial infarction and subsequent normal coronary arteriograms. Ann Intern Med 1974; 81:348.
553. Oakley DG, Oakley CM. Significance of abnormal electrocardiograms in highly trained athletes. Am J Cardiol 1982; 50:985.
554. Oka M and Angrist A. Histoenzymatic studies of vessels in hypertensive rats. Lab Invest 1967; 16:25.
555. Oliva PB, Breckinridge JC. Arteriographic evidence of coronary arterial spasm in acute myocardial infarction. Circulation 1977; 56(3):366.
556. Oliva PB, Hammill SC, Edwards WD. Cardiac rupture, a clinically predictable complication of acute myocardial infarction: report of 70 cases with clinicopathologic correlations. J Am Coll Cardiol 1993; 22:720.
557. Oliva PB, Potts DE, Plus RG. Coronary arterial spasm in Prinzmetal angina. N Engl J Med 1973; 288:745.
558. Oliver MF. Sudden cardiac death: lessons from epidemiology: summary of general discussion. J Am Coll Cardiol 1985; 5:155B.
559. Oparil S. Morphologic substrates of sudden death: Pathogenesis of ventricular hypertrophy. J Am Coll Cardiol 1985; 5:57B.
560. Opie L. Sudden death and sport. Lancet 1975; 1:263.
561. Opie LH. The mechanism of myocyte death in ischaemia. Eur Heart J 1993; 14 (Suppl G):31.
562. Osborn GR. The incubation period of coronary thrombosis. Butterworths, London, 1963.
563. Packer M. Sudden unexpected death in patients with congestive heart failure: a second frontier. Circulation 1985; 72:681.
564. Paessens R, Borchard F. Morphology of cardiac nerves in experimental infarction of rat hearts. I Fluorescence microscopical findings. Virchows Arch A Path Anat Histol 1980; 386:265.
565. Pagenstecher: Weiterer Beitrag zur Herzchirurgie. Die Unterbindung der verletzen Arteria Coronaria. Dtsch Med Wochenschr 4:56,1901
566. Pantano JA, Oriel RJ. Prevalence and nature of cardiac arrhythmias in apparently normal well-trained runners. Am Heart J 1982; 104:762.
567. Parmley LF, Mattingly TW, Manion WC. Penetrating wounds of the heart and aorta. Circulation 1958; 17:953.
568. Parmley WW. Factors causing arrhythmias in chronic congestive heart failure. Am Heart J 1987; 114:1267.
569. Parodi O, De Maria R, Oltrona L et al. Myocardial blood flow distribution in patients with ischemic heart disease or dilated cardiomyopathy undergoing heart transplantation. Circulation 1993; 88:509.
570. Parravicini C, Baroldi G, Gaiera G et al. Phenotype of intramyocardial leukocytic infiltrates in acquired immunodeficiency syndrome (AIDS): a postmortem immunohistochemical study in 34 consecutive cases. Mod Pathol 1991; 4:559.
571. Parums D, Mitchinson MJ. Demonstration of immunoglobulin in the neighbourhood of advanced atherosclerotic plaques. Atherosclerosis 1981; 38(1-2):211.
572. Pashkow F, Holland R, Brooks H. Early changes in contractility and coronary blood flow in the normal areas of the ischemic porcine heart. Am Heart J 1977; 93:349.
573. Pasterkamp G, Falk E, Woutman H et al. Techniques characterizing the coronary atherosclerotic plaque influence on clinical decision making? J Am Coll Cardiol 2000; 36:13.
574. Pasternac A, Tubau JF, Puddu PE et al. Increased plasma catecholamine levels in patients with symptomatic mitral valve prolapse. Am J Med 1982; 73:783.
575. Paterson JC. Capillary rupture with intimal hemorrhage as a causative factor in coronary thrombosis. Arch Path 1938:25:474.

576. Patterson E, Holland K, Eller BT et al. Ventricular fibrillation resulting from ischemia at a site remote from previous myocardial infarction. A conscious canine model of sudden coronary death. Am J Cardiol 1982; 50:1414.
577. Perper JA, Kuller LH, Cooper M. Arteriosclerosis of coronary arteries in sudden unexpected deaths. Circulation 1975; 51-52(Suppl 3):27.
578. Peters RW, Mitchell LB, Brooks MM et al. Circadian pattern of arrhythmic death in patients receiving encainide, flecainide or moricizine in the cardiac arrhythmia suppression trial (CAST). J Am Coll Cardiol 1994; 23:283.
579. Pethig K, Heublein B, Meliss RR et al. Volumetric remodeling of the proximal left coronary artery: early versus late after heart transplantation. Am J Coll Cardiol 1999; 34:197.
580. Piek JJ, Becker AE. Collateral blood supply to the myocardium at risk in human myocardial infarction: a quantitative postmortem assessment. J Am Coll Cardiol 1988; 11:1290.
581. Pierard LA, De Landsheere CM, Berthe C et al. Identification of viable myocardium by echocardiography during dobutamine infusion in patients with myocardial infarction after thrombolytic therapy: comparison with positron emission tomography. J Am Coll Cardiol 1990; 15:1021.
582. Piper C, Bilger J, Henrichs EM et al. Is myocardial $Na^{+}/Ca^{++}$ exchanger transcription a marker for different stages of myocardial dysfunction? Quantitative polymerase chain reaction of the messenger RNA in endomyocardial biopsies of patients with heart failure. J Am Coll Cardiol 2000; 36:233.
583. Pool J, Kunst K, Wermeskerken van JL. Two monitored cases of sudden death outside hospital. Brit Heart J 1978; 40:627.
584. Pool-Wilson PA. Relation of pathophysiologic mechanisms to outcome in heart failure. J Am Coll Cardiol 1993; 22 Supp A:A22-9.
585. Pratt CM, Francis MJ, Luck JC et al. Analysis of ambulatory electrocardiograms in 15 patients during spontaneous ventricular fibrillation with special reference to preceding arrhythmic events. J Am Coll Cardiol 1983; 2:789.
586. Proust F, Pujet JCP, Lubin S et al. Mort subite au cours d'un enregistrement electrocardiographique continu par la methode de Holter. Arch Mal Coeur 1981; 74:99.
587. Przyklenk K, Bauer B, Ovize M et al. Regional ischemic 'preconditioning' protects remote virgin myocardium from subsequent sustained coronary occlusion. Circulation 1993; 87:893.
588. Przyklenk K, Kloner RA. Superoxide dismutase plus catalase improve contractile function in the canine model of the stunned myocardium. Circ Res 1986; 58:148.
589. Pupita G, Maseri A, Kaski J C et al. Myocardial ischemia caused by distal coronary-artery constriction in stable angina pectoris. N Engl J Med 1990; 323:514.
590. Quillen JE, Rossen JD, Oskarsson HJ et al. Acute effect of cigarette smoking on the coronary circulation: constriction of epicardial and resistance vessels. J Am Coll Cardiol 1993; 22:642.
591. Quyyumi AA, Diodati JG, Lakatos E et al. Angiogenic effects of low molecular weight heparin in patients with stable coronary artery disease: a pilot study. J Am Coll Cardiol 1993; 22:635.
592. Raab W. Preventive Myocardiology, Fundamentals and Targets. In: Kugelmass NI, ed. Bannerstone Division of American Lectures in Living Chemistry. Springfield: CC Thomas Publisher, 1970.
593. Rabson MS, Helpern M. Sudden and unexpected death. II coronary atherosclerosis. Am Heart J 1948; 35:635.
594. Rafflenbeul W, Smith LR, Rogers WL et al. Quantitative coronary arteriography. Coronary anatomy of patients with unstable angina pectoris reexamined 1 year after optimal medical therapy. Am J Cardiol 1979; 43:699.
595. Rahimtoola SH. The hibernating myocardium. Am Heart J 1989; 117:211.
596. Rajanayagam SMA, Shou M, Thirumurti V et al. Intracoronary basic fibroblast growth factor enhances myocardial collateral perfusion in dogs. J Am Coll Cardiol 2000; 35:519.
597. Reagan TJ, Wu CF, Weisse AB et al. Acute myocardial infarction in toxic cadiomyopathy without coronary obstruction. Circulation 1975; 51:453.
598. Reduto LA, Freund GC, Gaeta JM et al. Coronary artery reperfusion in acute myocardial infarction:beneficial effects of intracoronary streptokinase on left ventricular salvage and performance. Am Heart J 1981; 102:1168.
599. Reichenbach DD, Benditt EP. Catecholamine and cardiomiopathy: the pathogenesis and potential importance of myofibrillar degeneration. Hum Path 1970; 1:125.

600. Reichenbach DD, Benditt EP. Myofibrillar degeneration: a common form of cardiac muscle injury. Ann NY Ac Sci 1969; 156:164.
601. Reichenbach DD, Moss NS, Meyer E. Pathology of the heart in sudden cardiac death. Am J Cardiol 1977; 39:865.
602. Reichenbach DD, Moss NS. Myocardial cell necrosis and sudden death in humans. Circulation 1975; 51-52 (Suppl 3):60.
603. Reimer KA, Jennings RB. The changing anatomic reference base of evolving myocardial infarction. Understimation of myocardial collateral blood flow and overestimation of experimental anatomic infarct size due to edema, hemorrhage and acute inflammation. Circulation 1979; 60:866.
604. Reimer KA, Lowe JE, Rasmussen MM et al. The wavefront phenomen of ischemic cell death. I Myocardial infarct size vs duration of coronary occlusion in dogs. Circulation 1977; 56:786.
605. Reimer KA, Rasmussen MM, Jennings RB. On the nature of protection by propranolol against myocardial necrosis after temporary coronary occlusion in dogs. Am J Cardiol 1976; 37:520.
606. Reiner L. Gross examination of the heart. In: Gould SE, ed. Pathology of the Heart and Blood Vessels. Springfield: Thomas CC, 1968:1136.
607. Rentrop KP, Cohen M, Blanke H et al. Changes in collateral channel filling immediatly after controlled coronary artery occlusion by an angioplasty balloon in human subjects. J Am Coll Cardiol 1985; 5:587.
608. Rentrop KP, Feit F, Sherman W et al. Late thrombolytic therapy preserves left ventricular function in patients with collateralized total coronary occlusion: primary end point findings of the second mount Sinai-New York University reperfusion trial. J Am Coll Cardiol 1989; 14:58.
609. Rentrop KP, Feit F, Sherman W et al. Serial angiographic assessment of coronary artery obstruction and collateral flow in acute myocardial infarction. Report from the second Mount Sinai-New Yok University reperfusion trial. Circulation 1989; 80:1166.
610. Rentrop KP, Thornton JC, Feit F et al. Determinants and protective potential of coronary arterial collaterals as assessed by an angioplasty model. Am J Cardiol 1988; 61:677.
611. Rentrop PK. Thrombi in acute coronary syndromes. Revisited and revised. Circulation 2000; 101:1619.
612. Rider AK, Bieber PP, Harrison DC. Atherosclerosis in long-term survivors of human cardiac allograft transplantation. Circulation 1972; 46 (Suppl 2):77.
613. Ridolfi RL, Hutchins GM. The relationship between coronary artery lesions and myocardial infarcts: ulceration of atherosclerotic plaques precipitating coronary thrombosis. Am Heart J 1977; 93:468.
614. Rigo P, Becker LC, Griffith LSC et al. Influence of coronary collateral vessels on the results of thallium-201 myocardial stress imaging. Am J Cardiol 1979; 44:452.
615. Roberts WC, Buja LM. The frequency and significance of coronary arterial thrombi and other observations in fatal acute myocardial infarction. Am J Med 1972; 52:425.
616. Roberts WC, Curry RC, Isner JM et al. Sudden death in Prinzmetal's angina with coronary spasm documented by angiography. Analysis of three necropsy patients. Am J Cardiol 1982; 50:203.
617. Roberts WC, Jones AA. Quantitation of coronary arterial narrowing at necropsy in sudden coronary death. Analysis of 31 patients and comparison with 25 control subjects. Am J Cardiol 1979; 44:39.
618. Roberts WC, Siegel RJ, Zipes DP. Origin of the right coronary artery from the left sinus of Valsalva and its functional consequences: analysis of 10 necropsy patients. Am J Cardiol 1982; 49:863.
619. Rodbard S. Physical factors in arterial sclerosis and stenosis. Angiology 1971; 22:267.
620. Rodbard S. Vascular modifications induced by flow. Am Heart J 1956; 51:926.
621. Rogers WJ, Hood WP, Mantle JA et al. Return of left ventricular function after reperfusion in patients with myocardial infarction: importance of subtotal stenoses or intact collaterals. Circulation 1984; 69:338.
622. Rona G, Chappel CI, Balazs T et al. An infart-like myocardial lesion and other toxic manifestations produced by isoproterenol in rat. Arch path 1959; 67:443.
623. Rose AG, Uys CJ. Pathology of cardiac transplantation. p. 1649 In: Silver MD ed. Cardiovascular Pathology. 2nd ed. New York: Churchill Livingstone, 1991:1649.
624. Rose AG, Viviers L, Odell JA. Pathology of chronic cardiac rejection: an analysis of the epicardial and intramyocardial coronary arteries and myocardial alterations in 43 humans allografts. Cardiovasc Pathol 1993; 2:7.

625. Rosenschein U, Ellis SG, Haudenschild CC et al. Comparison of histopathologic coronary lesions obtained from directional atherectomy in stable angina versus acute coronary syndromes. Am J Cardiol 1994; 73:508.
626. Ross R. The arterial wall and atherosclerosis. Ann Rev Med 1979; 30:1.
627. Ross R. The pathogenesis of atherosclerosis: a perspective for the 1990s. Nature 1993; 362:801.
628. Rossi L. Cardioneuropathy and extracardiac neural disease. J Am Coll Cardiol 1985; 5 (Suppl 6):66B.
629. Rossi L. Pathologic changes in the cardiac conduction and nerves system in sudden coronary death. Ann NY Acad Sci 1982; 382:50.
630. Rossi MA. Pathologic fibrosis and connective tissue matrix in left ventricular hypertrophy due to chronic arterial hypertension in humans. J Hypertens 1998; 16:1031.
631. Rozanski A, Blumenthal JA, Kaplan J. Impact of psychological factors on the pathogenesis of cardiovascular diseases and implication for therapy. Circulation 1999; 99:2196.
632. Ruberman W, Weinblatt E, Frank CW et al. Repeated 1 hour electrocardiographic monitoring of survivors of myocardial infarction at 6 month intervals: arrhythmia detection and relation to prognosis. Am J Cardiol 1981; 47:1197.
633. Ruffer MA. On arterial lesions found in Egyptian mummies. J Path Bact 1911; 15:453.
634. Rundqvist B, Elam M, Bergmann-Sverridottir Y et al. Increased cardiac adrenergic drive precedes generalized sympathetic activation in human heart failure. Circulation 1997; 95:169.
635. Sabbah HN, Goldberg AD, Schoels W et al. Spontaneous and inducible ventricular arrhythmias in a canine model of chronic heart failure: relation to haemodynamics and sympathoadrenergic activation. Eur Heart J 1992; 13:1562.
636. Saber RS, Edwards WD, Bailey KR et al. Coronary embolization after balloon angioplasty or thrombolytic therapy: an autopsy study of 32 cases. J Am Coll Cardiol 1993; 22:1283.
637. Sabia PJ, Powers ER, Jayaweera AR. Functional significance of collateral blood flow in patients with recent acute myocardial infarction. A study using myocardial contrast echocardiography. Circulation 1992; 85:2080.
638. Sabia PJ, Powers ER, Ragosta M et al. An association between collateral blood flow and myocardial viability in patients with recent myocardial infarction. N Engl J Med 1992; 327:1825.
639. Saffitz JE, Sazama K, Roberts WC. Amyloidosis limited to small arteries causing angina pectoris and sudden death. Am J Cardiol 1983; 51:1234.
640. Saito Y, Yasuno M, Ishida M et al. Importance of coronary collaterals for restoration of left ventricular function after intracoronary thrombolysis. Am J Cardiol 1985; 55:1259.
641. Sakama H, Globits S, O'Sullivan M et al. Breath-old MR measurements of blood flow velocity in internal mammary arteries and coronary artery bypass grafts. J Magn Reson Imaging 1996; 6:219.
642. Sakata K, Miura F, Sugino H et al. Assessment of regional sympathetic nerve activity in vasospastic angina: analysis of iodine-123-labeled metaiodobenzylguanidine scintigraphy. Am Heart J 133; 484.
643. Sakata K, Shirotani M, Yoshida H et al. Iodine-123 metaiodobenzylguanidine cardiac imaging to identify and localize vasospastic angina without significant coronary artery narrowing. Am J Coll Cardiol 1997; 30; 370.
644. Sakata Y, Kodama K, Kitanaza M et al. Different mechanism of ischemic adaptation to repeated coronary occlusions in patients with and without recruitable collateral circulation. J Am Coll Cardiol 1997; 30:1679.
645. Salomon V, Niemela M, Miettinen H et al. Relationship of socioeconomic status to the incidence and prehospital, 28 days and 1-year mortality rates of acute events in the Finmonica myocardial register study. Circulation 2000; 101:1913.
646. Sambuceti G, Giorgetti A, Consiglia L et al. Perfusion-contraction mismatch during inotropic stimulation in hibernating myocardium. J Nucl Med 1998; 39:396.
647. Sambuceti G, Giorgetti A, Corsiglia L et al. Perfusion-contraction mismatch during inotropic stimulation in hibernating myocardium. J Nucl Med 1998; 39:396.
648. Sambuceti G, Marzilli M, Maraccini P et al. Coronary vasoconstriciton during myocardial ischemia induced by rises in metabolic demand in patients with coronary artery disease. Circulation 1997; 95:2652.

649. Sambuceti G, Marzilli M, Maraccini P et al. Coronary vasoconstriction during myocardial ischemia induced by rises in metabolic demand in patients with coronary artery disease. Circulation 1997; 95:2652.
650. Sand NPR, Rehling M, Bagger JP et al. Functional significance of recruitable collaterals during temporary occlusion evaluated by $^{99m}$Tc-sestamibi single-photon emission computerized tomography. J Am Coll Cardiol 2000; 35:624.
651. Sangiorgi G, Rumberger JA, Severson A et al. Arterial calcification and not lumen stenosis is highly correlated with atherosclerotic plaque burden in human: a histologic study of 723 coronary artery segments using nondecalcifying method. J Am Coll Cardiol 1998; 31:126.
652. Saphir O, Kasner HT. An anatomical and experimental study of segmentation of the myocardium and its relationship to the intercalated disc. J Med Res 1924; 44:539.
653. Saphir O. Anatomical evidence of functional disorders of the heart. Arch Path 1933; 16:315.
654. Saram M. Ueber die azellulare Entstehung von Narben bei Durchblutunzgsstorungen in Herzmuskel. Beitr Pathol Anat Allg Pathol 1957; 118:275.
655. Sasaki H, Hoshi H, Hong YM et al. Purification of acidic fibroblast growth factor fro mboveine heart and its localization in the cardiac myocytes. J Biol Chem 1989; 264:17606.
656. Sasayama S, Masotoshi F. Recent insights into coroanry collateral circulation. Circulation 1992; 85:1197.
657. Schaeffers M, Lerch H, Wichter T et al. Cardiac sympathetic innervation in patients with idiopathic ventricular outflow tract tachycardia. J Am Coll Cardiol 1998; 32:181.
658. Schaper J, Borgers M, Schaper W. Ultrastructure of ischemia-induced changes in the precapillary anastomotic network of the heart. Am J Cardiol 1972; 29:851.
659. Schaper W, Pasyk S. Influence of collateral flow on the ischemic tolerance of the heart following acute and subacute coronary occlusion. Circulation 1976; 53:I57.
660. Scheel KW, Seavey E, Gaugl JF et al. Coronary and myocardial adaptations to high altitude in dogs. Am J Physiol 1990; 259 (Heart Circ Physiol 28) H1667.
661. Scheel KW, Williams SE. Hypertophy and coronary and collateral vascularity in dogs with severe chronic anemia. Am J Physiol 1985; 249 (Heart Circ Physiol 18) H103.
662. Schlesinger MJ, Reiner L. Focal myocytolysis of the heart. Am J Path 1955; 31:443.
663. Schlesinger MJ. An injection plus dissection study of coronary artery occlusions and anastomoses. Am Heart J 1938; 15:528.
664. Schlesinger MJ. Relation of anatomic pattern to pathological conditions of the coronary arteries. Arch Path 1940; 30:403.
665. Schofer J, Spielman R, Schuchert A et al. Iodine-123 meta-iodobenzylguanidine scintigraphy:a non invasive method to demonstrate myocardial adrenergic nervous system disintegrity in patients with idiopathic dilated cardiomyopathy. J Am Coll Cardiol 1988; 12:1252.
666. Schornagel HE. Intimal thickening in the coronary arteries in infants. Arch Path 1956; 62:427.
667. Schroder ES, Sirna SJ, Kieso RA et al. Sensitization of reperfused myocardium to subsequent coronary flow reductions. An extension of the concept of myocardial stunning. Circulation 1988; 78:717.
668. Schroeder JS, Berke DK, Graham AF et al. Arrhythmias after cardiac transplantation. Am J Cardiol 1974; 33:604.
669. Schwaiger M, Schelbert HR, Ellison D et al. Sustained regional abnormalities in cardiac metabolism after transient ischemia in the chronic dog model. J Am Coll Cardiol 1985; 6:336.
670. Schwartz CJ, Gerrity RG. Anatomical pathology of sudden unexpected cardiac death. Circulation 1975; 51-52(Suppl 3):18.
671. Schwartz CJ, Mitchell JRA. Cellular infiltration of the human arterial adventitia associated with atheromatous plaque. Circulation 1962; 26:73.
672. Schwartz ER, Schoendube FA, Kostin S et al. Prolonged myocardial hibernation exacerbates cardiomyocyte degeneration and impairs recovery of function after revascularization. J Am Coll Cardiol 1998; 31:118.
673. Schwartz H, Leiboff RH, Bren GB et al. Temporal evolution of the human coronary collateral circulation after myocardial infarction. J Am Coll Cardiol 1984; 4:1088.
674. Schwartz H, Leiboff RH, Katz KJ et al. Arteriographic predictors of spontaneous improvement in left ventricular function after myocardial infarction. Circulation 1985; 71:466.

675. Schwartz PJ, Billman GE, Stone HL. Autonomic mechanisms in ventricular fibrillation induced by myocardial ischemia during exercise in dogs with healed myocardial infarction. Circulation 1984; 69:790.
676. Schwartz PJ, La Rovere MT, Vanoli E. Autonomic nervous system and sudden cardiac death. Experimental basis and clinical observations for post-myocardial infarction risk stratification. Circulation 1992; 85 (Suppl 1):1-77.
677. Schwarz ER, Speakman MT, Patterson M et al. Evaluation of the effects of intramyocardial injection of DNA expressing vascular endothelial growth factor (VEGF) in a myocardial infarction model in the rat: angiogenesis and angioma formation. J Am Coll Cardiol 2000; 35:1323.
678. Schwarz F, Schaper J, Becker V et al. Coronary collateral vessels: their significance for left ventricular histologic structure. Am J Cardiol 1982; 49:291.
679. Scott RF, Briggs TS. Pathologic findings in pre-hospital deaths due to coronary atherosclerosis. Am J Cardiol 1972,29:782.
680. Serruys PW, Wijns W, Van den Brand M et al. Left ventricular performance, regional blood flow, wall motion, and lactate metabolism during transluminal angioplasty. Circulation 1984; 70:25.
681. Shan K, Bick RJ, Poindexter BJ et al. Relation to tissue Doppler derived myocardial velocities to myocardial structure and beta-adrenergic receptor density in humans. J Am Coll Cardiol 2000; 36:891.
682. Shanes JG, Ghali J, Billinghan ME et al. Interobserver variability in the pathologic interpretation of endomyocardial biopsy results. Circulation 1987; 75:401.
683. Sharma B, Asinger R, Francis G et al. Demonstration of exercise-induced painless myocardial ischemia in survivors of out-of-hospital ventricular fibrillation. Am J cardiol 1987; 59:740.
684. Shen AC, Jennings RB. Myocardial calcium and magnesium in acute ischemic injury. Am J Path 1972; 67:417.
685. Shen AC, Jennings RB. Unusual features of cell injury in cells on the edge of myocardial infarcts. Am J Path 1971; 62: abs.171.
686. Shephered JT. The heart as a sensory organ. J Am Coll Cardiol 1985; 5:83B.
687. Sherman CT, Litvack F, Grundfest WS et al. Demonstration of thrombus and complex atheroma by in-vivo angioscopy in patients with unstable angina pectoris. N Eng J Med 1986; 315:913.
688. Shibata M, Sakuma H, Isaka N et al. Assessment of coronary flow reserve with fast cinephase contrast magnetic resonance imaging: comparison with measurement by doppler guide wize. J Magn Reson Imaging 1999; 10:563.
689. Shizukuda Y, Buttrick PM, Geenen DL et al. Beta-adrenergic stimulation causes cardiocyte apoptosis: influence of tachycardia and hypertrophy. Am J Physiol 1998; 275(3 Pt 2):H961.
690. Shvalev VN, Virkhert AM, Stropus RA et al. Changes in neural and humoral mechanisms of the heart in sudden death due to myocardial abnormalities. J Am Coll Cardiol 1986; 8:55A.
691. Silver MD, Baroldi G, Mariani F. The relationship between acute occlusive coronary thrombi and myocardial infarction studies in 100 consecutive patients. Circulation 1980; 61:219.
692. Silver MD, Butany J, Chiasson. The pathology of myocardial infarction and its mechanical complications. In: David TE, ed. Mechanical complications of Myocardial Infarction. Austin: Landes Publishers 1993; 2:4.
693. Silver MD, Wigle ED, Trimble AS et al. Iatrogenic coronary ostial stenosis. Arch Path 1969; 88:73.
694. Silver MD. Medial hemorrhage and dissection in a coronary artery: an unusual cause of coronary occlusion. CMAJ 1968; 99:32.
695. Silver MM, Denic N, Smith CR. Development of the respiratory diaphragm in childhood: diaphragmatic contraction band necrosis in sudden death. Hum Pathol 1996; 27:57.
696. Silver MM, Freedom RM. Gross examination and structure of the heart. In: Silver MD ed. Cardiovascular Pathology. 2nd ed. New York: Churchill Livingston, 1991:1
697. Silver MM, Smith CR. Diaphragmatic contraction band necrosis in a perinatal and infantile autopsy population. Hum Pathol 1992; 23:817.
698. Sinapius O. Bezichungen zwischen Koronarthrombosen und Myokardinfarkten. Dtsch med Wschr 1972; 17:443.

699. Sinusas AJ, Hardin NJ, Clements JP et al. Pathoanatomic correlates of regional left ventricular wall motion assessed by equilibrium radionuclide angiocardiography: a postmortem correlation. Am J Cardiol 1984; 54:975.
700. Skinner JE. Regulation of cardiac vulnerability by the cerebral defense system. J Am Coll Cardiol 1985; 5:88B.
701. Smith SH, Kirling JK, Geer JC et al. Arteritis in cardiac rejection after transplantation. Am J Cardiol 1987; 59:1171.
702. Slades AKB, Saumarez RC, McKenna WJ. The arrhythmogenic substrato. Diagnostic and therapeutic implications in hypertrophic cardiomyopathy. Eur Heart J 1993; 14(Suppl E):84.
703. Snow PID, Jones AM, Daber KS. Coronary disease: a pathological study. Brit Heart J 1955; 17:503.
704. Soffer J, Dreifus LS, Michelson EL. Polimorphous ventricular tachycardia associated with normal and long Q-T intervals. Am J Cardiol 1982; 49:2021.
705. Sommers HM, Jennings RB. Ventricular fibrillation and myocardial necrosis after transient ischemia. Effect of treatment with oxygen, procainamide, reserpine and propranolol. Arch Int Med 1972; 129:780.
706. Spain DM, Bradess VA, Iral P et al. Intercoronary anastomotic channels and sudden unexpected death from advanced coronary atherosclerosis. Circulation 1963; 27:12.
707. Spain DM, Bradess VA. Postmortem studies on coronary atherosclerosis in one population group. Dis Chest 1959; 36:397.
708. Spain DM. Sudden death from coronary artery disese: survival time, frequency of thrombi and cigarette smoking. Am J Cardiol 1970; 25:129.
709. Spalteholz W, Hockrein M. Untersuchungen am Koronarsystem: die anatomische und funktionelle Beschaffenheit der Koronarterienwand. Arch Exp Path u Pharmakol 1931; 163:333.
710. Spalteholz W. Die Arterien der Herzwand. Anatomische Untersuchungen an Menschen und Tieren. S Hirzel, Leipsig, 1924
711. Speir E, Zhou YF, Lee M et al. Fibroblast growth factors are preent in adult cardiac myocytes, in vivo. Biochem Biophys Res Commun 1988; 157:1336.
712. Spiekerman RE, Brandenburg JT, Achor RWP et al. The spectrum of coronary heart disease in a Community of 30,000. a clinicopathologic study. Circulation 1962; 25:57.
713. Spinnler MT, Lombardi F, Moretti C et al. Evidence of functional alterations in sympathetic activity after myocardial infarction. Eur Heart J 1993; 14:1334.
714. Spiro D, Lattes RG, Wiener J. The cellular pathology of experimental hypertension. I. Hyperplastic arteriosclerosis. Am J Path 1965; 47:19.
715. Spiro D, Spotnitz H, Sonnenblick EH. The relation of cardiac fine structure to function. In: Gould SE, ed. Pathology of the Heart and Blood Vessels. Springfield: Charles C Thomas, 1968:131.
716. Staemmler M, Hertz. In: E. Kaufmann: Lehrbuch der speziellen pathologische Anatomie, De Gruyter W, Berlin, 1961:3.
717. Stahl LD, Aversano T, Ambrosio G, et al. Effect of repeated episodes of drug-induced ventricular dyskinesia on subsequent regional function in the dog: comparison with myocardial stunning produced by repeated coronary occlusions. J Am Coll Cardiol 1987; 9:1339.
718. Stanton MS, Tuli MM, Radtke NL et al. Regional sympathetic denervation after myocardial infarction in humans detected noninvasively using I-123-metaiodobenzyeguanidine. J Am Coll Cardiol 1989; 14:1519.
719. Stevensen MJ, Raffael DM, Allman KC et al. Cardiac sympathetic dysinnervation in diabetes. Implication for enhanced cardiovascular risk. Circulation 1998; 98:961.
720. Still WJS. The pathogenesis of the intimal thickenings produced by hypertension in large arteries in the rat. Lab Invest 1968; 19:84.
721. Stinson EB, Billingham ME. Correlative study of regional left ventricular histology and contractile function. Am J Cardiol 1977; 39:378.
722. Strong JP, Solberg LA Restrepo C. Atherosclerosis in persons with coronary heart disease. Lab Invest 1968; 18:527.
723. Sugishita Y, Koseki S, Matsuda M et al. Myocardial mechanics of athletic hearts in comparison with diseased hearts. Am Heart J 1983; 105:273.
724. Summers WK, Jamison RL. The no-reflow phenomenon in renal ischemia. Lab Invest 1971; 25:635.

725. Sun SC, Burch GE, De Pasquale NP. Histochemical and electron microscopy study of heart muscle after beta-adrenergic blockade. Am Heart J 1967; 74:340.

726. Surawicz B. Ventricular fibrllation. J Am Coll Cardiol 1985; 5(Suppl6):42B.

727. Szakacs JE, Cannon A. L-norepinephrine myocarditis. Am J Clin Pathol 1958; 30:425.

728. Szakacs JE, Dimmette RM, Gowart EC. Pathologic implication of the catecholamines, epinephrine and norepinephrine. U.S. Armed Forces Med J 1959; 10:908.

729. Takano H, Nakamua T, Satou T et al. Regional myocardial sympathetic denervation in patients with coronary spasm. Am J Cardiol 1997; 75:324.

730. Takeda A, Sakata A, Takeda N. Image and DNA analysis of hypertrophic myocytes in hypertensive heart disease and hypertrophic cardiomyopathy. Analyt Quantit Cytol Histol 1999; 21:454.

731. Taki J, Yasuda T, Gold HK et al. Characteristic of transient left ventricular dysfunction detected by ambulatory left ventricular function monitoring device in patients with coronary artery disease (abstract). Circulation 1987; 76 (Suppl 4):IV 366.

732. Taylor AJ, Rogan KM, Virmani R. Sudden cardiac death associated with isolated congenital coronary artery anomalies. J Am Coll Cardiol 1992; 20:640.

733. Taylor AL, Murphree S, Buja ML et al. Segmental systolic response to brief ischemia and reperfusion in the hypertophied canine left ventricle. J Am Coll Cardiol 1992; 20:994.

734. Teare D. Asymmetrical hypertrophy of the heart in young adults. Br Heart J 1958; 20:1.

735. Tennant R, Grayzel DM, Sutherland FA et al. Studies on experimental coronary occlusion. Chemical and anatomical changes in the myocardium after coronary ligation. Am Heart J1936; 12:168.

736. Tennant R, Wiggers CJ. The effect of the coronary occlusion on myocardial contraction. Am J Physiol 1935; 112:351.

737. Thiene G, Gambino A, Corrado D et al.The pathological spectrum underlying sudden death in athletes. New Trends in Arrhythmias 1985; 3:323.

738. Thiene G, Nava A, Corrad D et al. Right ventricular cardiomyopathy and sudden death in young people. N Engl J Med 1988; 318:129.

739. Thiene G, Pennelli N, Rossi L. Cardiac conduction system abnormalities as a possible cause of sudden death in young athletes. Hum Pathol 1983; 14:709.

740. Thompson PD, Stern MP, Williams P et al. Death during jogging or running: a study of 18 cases.JAMA 1979; 242:1265.

741. Thomson JG. Production of severe atheroma in a transplanted human heart. Lancet 1969; 2:1088.

742. Tillmanns H, Ikeda S, Hansen H et al. Microcirculation in the ventricle of the dog and turtle. Circ Res 1974; 34:561.

743. Titus JL, Oxman HA, Nobrega FT et al. Sudden unexpected death as the initial manifestation of ischemic heart disease. Clinical and pathologic observations. Am J Cardiol 1970; 26:662.

744. Todd GL, Baroldi G, Pieper GM et al. Experimental catecholamine-induced myocardial necrosis. I. Morphology, quantification and regional distribution of acute contraction band lesions. J Mol Cell Cardiol 1985; 17:317.

745. Todd GL, Baroldi G, Pieper GM et al. Experimental catecholamine-induced myocardial necrosis. II Temporal development of isoproterenol-induced contraction band lesions correlated with ECG, hemodynamic and biochemical changes. J Mol Cell Cardiol 1985; 17:647.

746. Tomoike H, Ross Jr J, Franklin D et al. Improvement by propranolol of regional myocardial dysfunction and abnormal coronary flow pattern in conscious dogs with coronary narrowing. Am J Cardiol 1978; 41:689.

747. Topol EJ, Ellis SG. Coronary collaterals revisited - Accessory pathway to myocardial preservation during infarction. Circulation 1991; 83:1084.

748. Topol EJ, Yadav JS. Recognition of the importance of embolization in atherosclerotic vascular disease. Circulation 2000; 101:570.

749. Tsung SH, Huang TY, Chang HH. Sudden death in young athletes. Arch Pathol Lab Med 1982; 106:168.

750. Turitto G, Risa AL, Zanchi E et al. The signal-averaged electrocardiogram and ventricular arrhythmias after thrombolysis for acute myocardial infarction. J Am Coll Cardiol 1990; 15:1270.

751. Udelson JE, Coleman PS, Metherall J et al. Predicting recovery of severe regional ventricular dysfunction. Comparison of resting scintigraphy with $^{201}$TL and $^{99m}$Tc-sestamibi. Circulation 1994; 89:2552-2561.

752. Unger EF, Sheffield CD, Epstein SE. Creation of anastomoses between an extracardiac artery and the coronary circulation. Circulation 1990; 82:1449.
753. Ungerer M, Hartmann F, Karoglan et al. Regional in vivo and in vitro characterization of autonomic innervation in cardiomyopathic human heart. Circulation 1998; 97:174-180.
754. Uren NG, Marracini P, Gistri R et al. Altered coronary vasodilator reserve and metabolism in myocardium subtended by normal arteries in patients with coronary artery disease. J Am Coll Cardiol 1993; 22:650.
755. Vakeva AP, Azah A, Rollins SA et al. Myocardial infarction and apoptosis after myocardial ischemia and reperfusion. Role of the terminal complements and inhibition by anti-C5 therapy. Circulation 1998; 97:2259.
756. Val-Mejias J, Lee WK, Weisse AB et al. Left ventricular performance during and after sickle cell crisis. Am Heart J 1974; 97:585.
757. Vanderwee MA, Humphrey SM, Gavin JB et al. Changes in the contractile state: fine structure and metabolism of cardiac muscle cells during the development of rigor mortis. Virchows Arch (Cell Pathol) 1981; 35:159.
758. Vanoverschelde JJJ, Wijns W, Depré C et al. Mechanisms of chronic regional postischemic dysfunction in humans. New insights from the study of noninfarcted collateral-dependent myocardium. Circulation 1993; 87:1513.
759. Vassable M. On the mechanism underlying cardiac standstill: factors determining success or failure of escape pacemakers in the heart. J Am Coll Cardiol 1985; 5(Suppl6):35B.
760. Velican C, Velican D. Natural history of coronary atherosclerosis. CRC Press Inc, Boca Raton, Florida,1989.
761. Vincent MG, Anderson JL, Marshall HW. Coronary spasm producing coronary thrombosis and myocardial infarction. N Engl J Med 1983; 309:220.
762. Virmani N, Atkinson JB, Forman MB. Aortocoronary bypass grafts and extracardiac conduits. In: Silver MD, ed. Cardiovascular Pathology. 2nd ed. New York: Churchill Livingstone, 1991:1607.
763. Virmani R, Robinowitz M, McAllister Jr HA. Nontraumatic death in joggers: a series of 30 patients at autopsy. Am J Med 1982; 72:874.
764. Vismara LA, Miller RR, Price JE et al. Improved longevity due to reduction of sudden death by aortocoronary bypass in coronary atherosclerosis. Prospective evaluation of medical versus surgical therapy in matched patients with multivessel disease. Am J Cardiol 1977; 39:919.
765. Vita JA, Treasure CB, Ganz P et al. Control of shear stress in the epicardial coronary arteries of humans: impairment by atherosclerosis. J Am Coll Cardiol 1989; 14:1193.
766. Vlay SC, Burger L, Vlay LC et al. Prediction of sudden cardiac arrest: risk stratification by anatomic substrate. Am Heart J 1993; 126:807.
767. Vlodaver Z, Edwards JE. Pathology of coronary atherosclerosis. Progr Cardiov Dis 1971; 14:256.
768. Vlodaver Z, Kahn HA, Neufeld HN. The coronary arteries in early life in three different ethnic groups. Circulation 1969; 39:541.
769. Vlodaver Z, Neufeld HM. The musculo-elastic layer in the coronary arteries. A histological and hemodynamic concept. Vasc Dis 1967; 4:136.
770. Vlodaver Z, Neufeld HN. The coronary arteries in coarctation of the aorta. Circulation 1968; 37:449.
771. Vogt A, Von Essen R, Tebbe U et al. Impact of early perfusion status of the infarct-artery on short-term mortality after thrombolysis for acute myocardial infarction: retrospective analysis of four German multicenter studies. J Am Coll Cardiol 1993; 21:1391.
772. Voigt J, Agdal N. Lipomatous infiltration of the heart. Arch Pathol Lab Med 1982; 106:497.
773. Vongpatanasin W, Mansour Y, Chavoshan B et al. Cocaine stimulates the human cardiovascular system via a central mechanism of action. Circulation 1999; 100:497.
774. Wal van der AC, Becker AE, Loos van der CM, Das PK. Site of intimal rupture or erosion of thrombosed coronary atherosclerotic plaques is characterized by an inflammatory process irrespective of the dominant plaque morphology. Circulation 1994; 89:36.
775. Wal van der AC, Das PK, Berg van der DB et al. Atherosclerotic lesions in humans. In situ immunophenotypic analysis suggesting an immune mediated response. Lab Invest 1989; 61; 166.
776. Waller BF, Pinkerton CA. "Cutters, scoopers, shavers and scrapers": The importance of atherectomy devices and clinical relevance of tissue removed. J Am Coll Cardiol 1990; 15:426.

777. Waller BF, Roberts WC. Sudden death while running in conditioned runners aged 40 years or over. Am J Cardiol 1980; 45:1292.
778. Waller BF, Rothbaum DA, Pinkerton CA et al. Status of myocardium and infarcted-related artery in 19 necropsy patients with acute recanalization using pharmacologic (streptokinase, r-tissue plasminogen activator) mechanical (percutaneous transluminal coronary angioplasty) or combined types of reperfusion therapy. J Am Coll Cardiol 1987; 9:785.
779. Waller BF, Rothbaum DA, Pinkerton CA et al. Status of the myocardium and infarct-related coronary artery in 19 necropsy patients with acute recanalization using pharmacologic (streptokinase, r-tissue plasminogen activator), mechanical (percutaneous transluminal coronary angioplasty) or combined types of reperfusion therapy. J Am Coll Cardiol 1987; 9:785.
780. Waller BF. Pathology of new cardiovascular interventional procedures. In: Silver MD, ed. Cardiovascular Pathology. 2nd ed. New York: Churchill Livingstone, 1991:1683.
781. Wanibuchi H, Veda M, Dingermans KP et al. The response to percutaneous transluminal coronary angioplasty: an ultrastructural study of smooth muscle cells and endothelial cells. Cardiovasc Pathol 1992; 1:295.
782. Ware JA, Heistad DD. Platelet-endothelium interactions. N Engl J Med 1993; 328:628.
783. Warnes CA, Roberts WC. Sudden coronary death: comparison of patients with to those without coronary thrombus at necroscopy. Am J Cardiol 1984; 54:1206.
784. Warren JV. Di si dolce morte. It may be safer to be dead than alive. Circulation 1974; 50:415.
785. Wartman WB. Occlusion of the coronary arteries by hemorrhage into their walls. Am Heart J 1938; 15:459.
786. Wearn JT. The role of the Thebesian vessels in the circulation of the heart. J Exp Med 1928; 47:293.
787. Weber KT. Cardiac interstitium in health and disease: the fibrillar collagen network. J Am Coll Cardiol 1989; 13:1637.
788. Weber KT. Cardiac interstitium in health and disease: the fibrillar collagen network. J Am Coll Cardiol 1989; 13:1637.
789. Webster MW, Chesebro JH, Smith HC et al. Myocardial infarction and coronary artery occlusion: a prospective 5-year angiographic study.J Am Coll Cardiol 1990; 15:218A.
790. Weiner HL, Swain L. Acidic fibroblast growth factor in RNA es expressed by cardiac myocytes in culture and the protein is localized to the extracellular matrix. Proc Natl Acad Sci USA 1989; 86:2683.
791. Weiss DL, Subswicz B, Rubenstein I. Myocardial lesions of calcium deficiency causing irreversible myocardial failure. Am J Path 1966; 48:653.
792. Weisse AB, Lehan PH, Ettinger PO et al. The fate of experimentally induced coronary thrombosis. Am J Cardiol 1969; 23:229.
793. Wenger NK, Bauer S. Coronary embolism. Review of the literature and presentation of fifteen cases. Am J Med 1958; 25:549.
794. White M, Wiechmann RJ, Roden RL et al. Cardiac b-adrenergic neuroeffector systems in acute myocardial dysfunction related to brain injury. Evidence for catecholamine-mediated myocardial damage. Circulation 1995; 92:2183.
795. White PD, Boursy AV. Translation of De Subitaneis Mortibus (on sudden deaths). New York: St. John's University Press, 1970.
796. WHO Report: Scientific group on the pathological diagnosis of acute ischemic heart disease. Geneve, 1969.
797. Wichter T, Hindrcks G, Lerch H et al. Regional myocardial sympathetic dysinnervation in arrhythmogenic right ventricular cardiomyopathy. An analysis using I123 meta-iodobenzylguanidine scintigraphy. Circulation 1994; 89:667.
798. Wichter T, Schafers M, Rhodes CG et al. Abnormalities of cardiac sympathetic innervation in arrhythmogenic right ventricular cardiomyopathy. Quantitative assessment of presynaptic norepinephrine reuptake and postsynaptic b-adrenergic receptor density with positron emission tomography. Circulation 2000; 101:1552.
799. Wight TN, Curwen KD, Litrenta MM et al. Effect of endothelium on glycosaminoglycan accumulation in injured rabbit aorta. Am J Path 1983; 113:156.

800. Wijns W, Serruys PW, Slager CJ et al. Effect of coronary occlusion during percutaneous transluminal angioplasty in humans on left ventricular chamber stiffness and regional diastolic pressure-radius relations. J Am Coll Cardiol 1986; 3:455.
801. Wikstrand J, Kendall M. The role of beta receptor blockade in preventing sudden death. Eur Heart J 1992; 13 (Suppl D):111.
802. Wikstrand J, Warnold I, Olsson G et al. Primary prevention with metopolol in patients with hypertension. Mortality results from the MAPHY study. JAMA 1988; 259:1976.
803. William SR. Apoptosis and heart failure. N Engl J Med 1999; 341:759.
804. Williams DD, Amsterdam EA, Miller RR et al. Functional significance of coronary collateral vessels in patients with acute myocardial infarction: relation to pump performance, cardiogenic shock and survival. Am J Cardiol 1976; 37:345,
805. Willich SN, Maclure M, Mittleman M et al. Sudden cardiac death. Support for a role of triggering in causation. Circulation 1993; 87:1442.
806. Wilson JB. An introduction to scientific research. New York: McGraw-Hill, 1952:41.
807. Wilson RF, Laxson DD, Christenen RV et al. Regional differences in sympathetic reinnervation after human orthotopic cardiac transplantation. Circulation 1993; 88:165.
808. Wissler RW. Current status of regression studies. In: Paoletti R, Gotto Jr AM eds. Atherosclerosis Review. Vol 3. New York: Raven Press, 1978:213.
809. Wissler RW. The arterial medial cell, smooth muscle of multifunctional mesenchyme? Circulation 1967; 36:1.
810. Witzer DJ, Kaye MP. Myocardial ultrastructural changes induced by administration of nerve growth factors. Surgery forum 1976; 27:295.
811. Wolinsky H. Effects of estrogen and progestogen treatment on the response of the aorta of male rats to hypertension. Morphological studies. Circul Res 1972; 30:341.
812. Wolkoff K. Ueber die Atherosklerose der Coronarterien des Herzens . Beitr Path Anat 1929; 82:555.
813. Yamagashi M, Miyatake K, Tamai J et al. Intravascular ultra sound detection of atherosclerosis at the site of focal vasospasm in angiographically normal or minimally narrowed coronary segments. J Am Coll Cardiol 1994; 23:352.
814. Yamamoto H, Tomoike H, Shimokawa H et al. Development of collateral function with repetitive coronary occlusion in a canine model reduces myocardial reactive hyperemia in the absence of significant coronary stenosis. Circ Res 1984; 55:623.
815. Yamori Y, Mano M, Nara Y et al. Cathecolamine-induced polyploidization in vascular smooth muscle cells. Circulation 1987; 75 (suppl 1)1-92.
816. Yoshioka K, Gao OW, Chin M et al. Heterogeneous sympathetic innervation influences local myocardial repolarization in normally perfused rabbit hearts. Circulation 1987; 101:1060.
817. Yu Q, Zhao G, Marian AJ. In vivo short-term expression of a hypertrophic cardiomyopathy mutation in adult rabbit myocardium: myofibrillar incorporation without early disarray. Proceedings Ass Am Physicians 1999; 111:45.
818. Zahn R, Schiele R, Schneider S et al. Decreasing hospital mortality between 1994 and 1998 in patients with acute myocardial infarction treated with primary angioplasty but not in patients treated with intravenous thromnbolysis. J Am Coll Cardiol 2000; 36:2064.
819. Zamir M, Silver MD. Vasculature in the walls of human coronary arteries. Arch Pathol Lab Med 1985; 109:659.
820. Zerbini EJ. Coronary ligation in wounds of the heart. Report of a case in which ligation of the anterior descending branch of the left coronary artery was followed by complete recovery. J Thorac Surg 1943; 12:642.
821. Zhao M, Zhang H, Robinson TF et al. Profound structural alterations of the extracellular collagen matrix in postischemic dysfunctional ("stunned") but viable myocardium. J Am Coll Cardiol 1987; 10:1322.
822. Zimmermann ANE, Daems W, Hulsmann WC et al. Morphopathological changes of heart muscle caused by successive perfusion with calcium-free and calcium containing solutions (calcium paradox). Cardiov Res 1967; 1:201.
823. Zipes DP. Electrophysiological mechanism involved in ventricular fibrillation. Circulation 1975; 51-52(Suppl 3):3-120.

# Index